W0230078

Animal Sensitivity
and Behaviour

Animal Sensitivity and Behaviour

Hubert Lincoln

Editor

KOROS PRESS LIMITED
London, UK

Animal Sensitivity and Behaviour

© 2012
Printed in 2017 for Sale in the Indian Subcontinent

Published by
Koros Press Limited
3 The Pines, Rubery B45 9FF, Rednal,
Birmingham, United Kingdom

Tel.: +44-7826-930152
Email: info@korospress.com
www.korospress.com

ISBN: 978-1-78163-136-2

Editor: Hubert Lincoln

Printed in UK

British Library Cataloguing in Publication Data
A CIP record for this book is available from the British Library

10 9 8 7 6 5 4 3 2 1

No part of this publication may be reproduced, stored in a retrieval system or transmitted in any form or by any means, electronic, mechanical, photocopying, recording, scanning or otherwise without prior written permission of the publisher.

Reasonable efforts have been made to publish reliable data and information, but the authors, editors, and the publisher cannot assume responsibility for the legality of all materials or the consequences of their use. The authors, editors, and the publisher have attempted to trace the copyright holders of all materials in this publication and express regret to copyright holders if permission to publish has not been obtained. If any copyright material has not been acknowledged, let us know so we may rectify in any future reprint.

Exclusively distributed by CBS Publishers & Distributors Pvt. Ltd.

Sales & Distribution Rights only for India, Pakistan, Bangladesh, Sri Lanka, Nepal and Bhutan.This book is not to be sold outside these territories.

Contents

Preface

Much of the work in ethology revolves around problems in animal communication. In fact, this is the work for which von Frisch is best known. He teased apart the "dance language" of bees. Worker bees dance to communicate the distance, direction, and quality of a feeding place they have found while foraging around the hive. The dance is performed on the vertical comb of the hive. If the feeding place is directly towards the sun, the bees dance up. If the feeding place is directly away from the sun, they will dance down. As the sun moves across the sky, the bees adjust their dance so that the direction is always up-to-date. If, for example, a bee finds food at sunrise and indicates that it is located directly towards the sun by the time the sun sets, the bee will be dancing down. Thus, bees have an internal biological clock that tells them where the sun should be in the sky without them having to go out and check. Distance to the feeding place is communicated by the number of waggles at the center part of the dance. More waggles indicate greater distance. Sites with lots of high quality food are communicated by the vigour of the dance. In addition, other workers paying attention to the dancer touch their sister and can determine what type of flower she visited by the pollen stuck on her hairs. A second dance, a round dance, is used when the food source is very close to the hive.

With good vision, it became possible for many animals to use a new channel for communication. Among those spiders that have good vision, courtship dances perfumed by male spiders help to get the female in the mood. Other animals have more spectacular, albeit still silly, courtship dances. Competition among males for the most impressive displays to attract females has led to ornaments, colours, badges, and other decorations to attract females and/or ward off other male competitors. The red epaulets of the redwing blackbird, for example, are necessary for male birds to defend their territory. If the epaulets are painted black, the male can no longer chase off competing males. Among other animals, the need to impress females has gone to an extreme. The male peacock, for example, is required to put up with his ungainly and brightly-coloured tail and accessory plumage to impress females. This is despite the fact that these feathers make him more susceptible to predation and are metabolically expensive to construct. The Irish elk is now extinct, and has been since the end of last ice age. During the ice age, competition among males and selection by females resulted in huge antlers used for display and defence. When the glaciers retreated, forests grew in their place, and the huge antlers became a liability since the males couldn't move among the densely-packed trees without catching their heads. This ultimately lead to the extinction of the species. Generally, sexual selection leads males to extremes and is driven entirely by female preferences and competition among males for mates. This has produced some extremely silly looking and conspicuous males over evolutionary history.

The book provides a useful framework from which the pragmatics of zoology can be understood, and being quite comprehensive, should prove to be source.

—Editor

Chapter 1

Anatomy and Physiology of Animals
Main Organs

The Cell

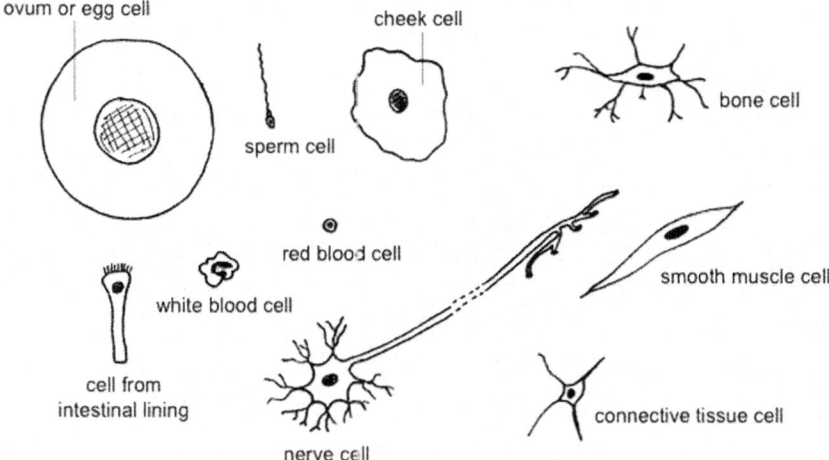

Figure 1: A variety of animal cells

The cell is the basic building block of living organisms. Bacteria and the parasite that causes malaria consist of single cells, while plants and animals are made up of trillions of cells. Most cells are spherical or cube shaped but some are a range of different shapes.

Most cells are so small that a microscope is needed to see them, although a few cells, e.g. the ostrich's egg, are so large that they could make a meal for several people.

A normal cell is about 0.02 of a millimetre (0.02mm) in diametre. (Small distances like this are normally expressed in micrometres or microns (îm). Note there are 1000 îms in every mm).

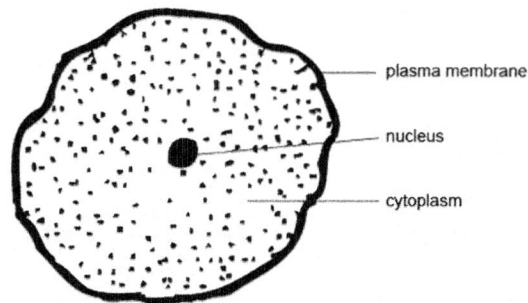

Figure 2: An animal cell

When you look at a typical animal cell with a light microscope it seems quite simple with only a few structures visible. Three main parts can be seen:

- an outer cell wall or plasma membrane,
- an inner region called the cytoplasm and
- the nucleus

However, when you use an electron microscope to increase the magnification many thousands of times you see that these seemingly simple structures are incredibly complex, each with its own specialised function. For example the plasma membrane is seen to be a double layer and the cytoplasm contains many special structures called organelles (meaning little organs) which are described below. A drawing of the cell as seen with an electron microscope.

The Plasma Membrane

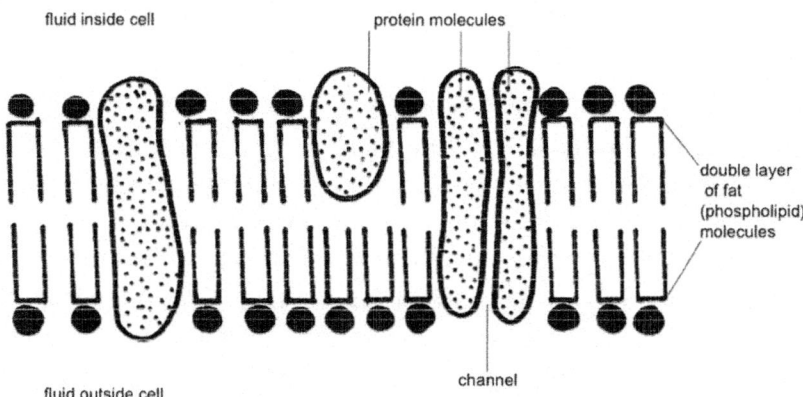

Figure 3: The structure of the plasma membrane

The thin plasma membrane surrounds the cell, separating its contents from the surroundings and controlling what enters and leaves the cell. The plasma membrane is composed of two main molecules, fats (in fact phospholipids) and proteins. The fats are arranged in a double layer with the large protein molecules dotted about in the membrane. Some of the protein molecules form tiny channels in the membrane while others help transport substances from one side of the membrane to the other.

How Substances Move Across the Plasma Membrane

Substances need to pass through the membrane to enter or leave the cell and they do so in a number of ways. Some of these processes require no energy i.e. they are passive, while others require energy i.e. they are active.

Passive processes include: a) diffusion and b) osmosis, while active processes include: c) active transport, d) phagocytosis, e) pinocytosis and f) exocytosis. These will be described below.

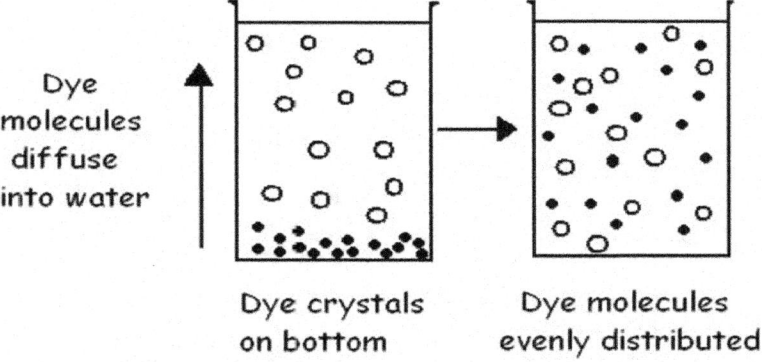

Dye molecules diffuse into water

Dye crystals on bottom

Dye molecules evenly distributed

Figure 4: Diffusion in a liquid

Diffusion: Although you may not know it, you are already familiar with the process of diffusion. It is diffusion that causes a smell (expensive perfume or smelly socks) in one part of the room to gradually move through the room so it can be smelt on the other side. Diffusion occurs in the air and in liquids.

What happens when a few crystals of a dark purple dye called potassium permanganate are dropped into a beaker of water. The dye molecules diffuse into the water moving from high to low concentrations so they become evenly distributed throughout the beaker.

In the body, diffusion causes molecules that are in a high concentration on one side of the cell membrane to move across the membrane until they are present in equal concentrations on both sides. It takes place because all molecules have an inbuilt vibration that causes them to move and collide until they are evenly distributed. It is an absolutely natural process that requires no added energy.

Small molecules like oxygen, carbon dioxide, water and ammonia as well as fats, diffuse directly through the double fat layer of the membrane. The small molecules named above as well as a variety of charged particles (ions) also diffuse through the protein-lined channels. Larger molecules like glucose attach to a carrier molecule that aids their diffusion through the membrane. This is called facilitated diffusion.

In the animal's body diffusion is important for moving oxygen and carbon dioxide between the lungs and the blood, for moving digested food molecules from the gut into the blood and for the removal of waste products from the cell.

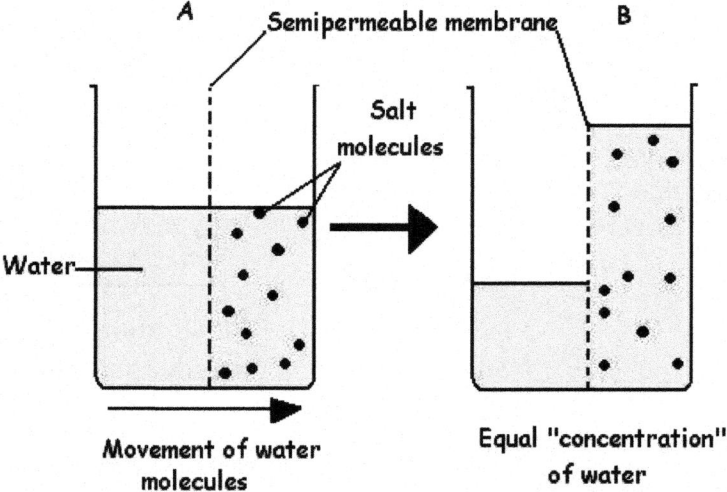

Figure 5: Osmosis

Osmosis: Although the word may be unfamiliar, you are almost certainly acquainted with the effects of osmosis. It is osmosis that plumps out dried fruit when you soak it before making a fruit cake or makes that wizened old carrot look almost like new when you soak it in water. Osmosis is in fact the diffusion of water across a membrane that allows water across but not larger molecules. This kind of membrane is called a semi-permeable membrane.

It shows a container divided into two parts by an artificial semi-permeable membrane. Water is poured into one part while a solution containing salt is poured into the other part. Water can cross the membrane but the salt cannot. The water crosses the semi-permeable membrane by diffusion until there is an equal amount of water on both sides of the membrane. The effect of this would be to make the salt solution more diluted and cause the level of the liquid in the right-hand side of the container to rise so it looked like side B. This movement of water across the semi-permeable membrane is called osmosis. It is a completely natural process that requires no outside energy.

Although it would be difficult to do in practice, imagine that you could now take a plunger and push down on the fluid in the right-hand side of container B so that it flowed back across the semi-permeable membrane until the level of fluid on both sides was equal again. If you could measure the pressure required to do this, this would be equal to the osmotic pressure of the salt solution. (This is a rather advanced concept at this stage but you will meet this term again when you study fluid balance later in the course).

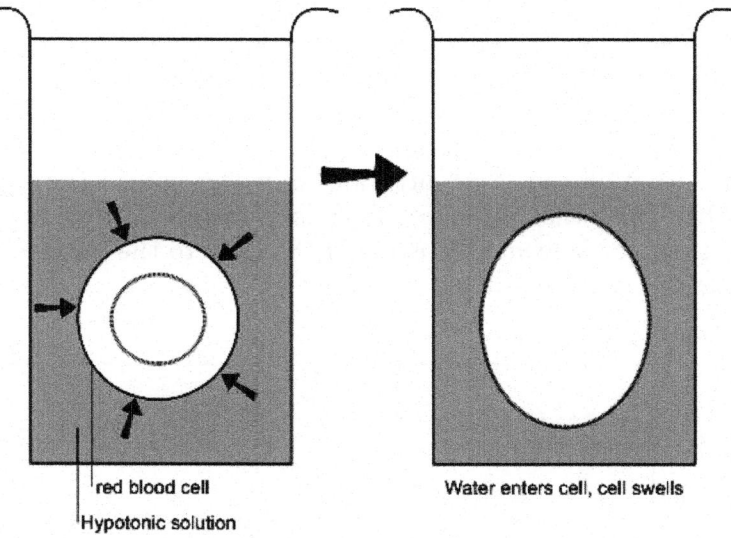

red blood cell

Hypotonic solution

Water enters cell, cell swells

Figure 6: Osmosis in red cells placed in a hypotonic solution

The plasma membrane of cells acts as a semi-permeable membrane. If red blood cells, for example, are placed in water, the water crosses the membrane to make the amount of water on both sides of it equal. This means that the water moves into the cell causing it to swell. This can occur to such an extent that the cell actually bursts to release its contents. This bursting of red blood cells is called haemolysis. In a situation such as this when the solution on one side of a semi-permeable membrane has a lower concentration than that on the other side, the first solution is said to be hypotonic to the second.

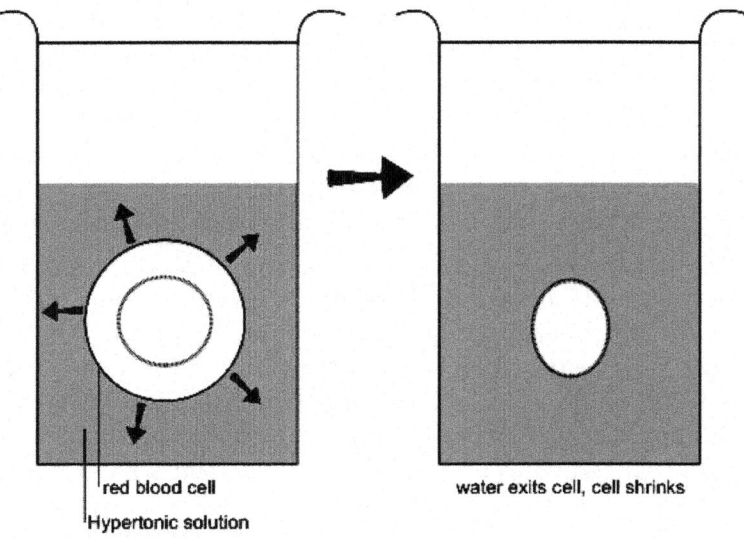

Figure 7: Osmosis in red cells placed in a hypertonic solution

Now think what would happen if red blood cells were placed in a salt solution that has a higher salt concentration than the solution within the cells. Such a bathing solution is called a hypertonic solution. In this situation the "concentration" of water within the cells would be higher than that outside the cells. Osmosis (diffusion of water) would then occur from the inside of the cells to the outside solution, causing the cells to shrink.

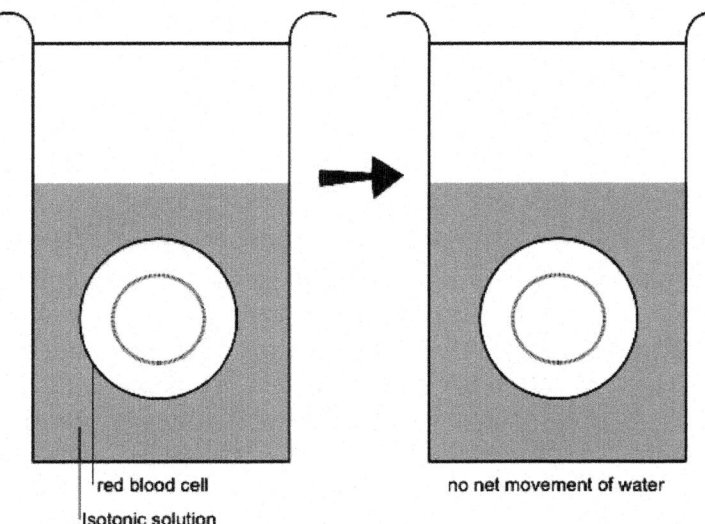

Figure 8: Red cells placed in an isotonic solution

A solution that contains 0.9% salt has the same concentration as body fluids and the solution within red cells. Cells placed in such a solution would neither swell nor shrink. This solution is called an isotonic solution. This strength of salt solution is often called normal saline and is used when replacing an animal's body fluids or when cells like red blood cells have to be suspended in fluid. Remember - osmosis is a special kind of diffusion. It is the diffusion of water molecules across a semi-permeable membrane. It is a completely passive process and requires no energy.

Sometimes it is difficult to remember which way the water molecules move. Although it is not strictly true in a biological sense, many students use the phrase "SALT SUCKS" to help them remember which way water moves across the membrane when there are two solutions of different salt concentrations on either side. As we have seen water moves in and out of the cell by osmosis. All water movement from the intestine into the blood system and between the blood capillaries and the fluid around the cells (tissue or extra cellular fluid) takes place by osmosis. Osmosis is also important in the production of concentrated urine by the kidney.

Active Transport: When a substance is transported from a low concentration to a high concentration i.e. uphill against the concentration gradient, energy has to be used. This is called active transport.

Active transport is important in maintaining different concentrations of the ions sodium and potassium on either side of the nerve cell membrane. It is also important for removing valuable molecules such as glucose, amino acids and sodium ions from the urine.

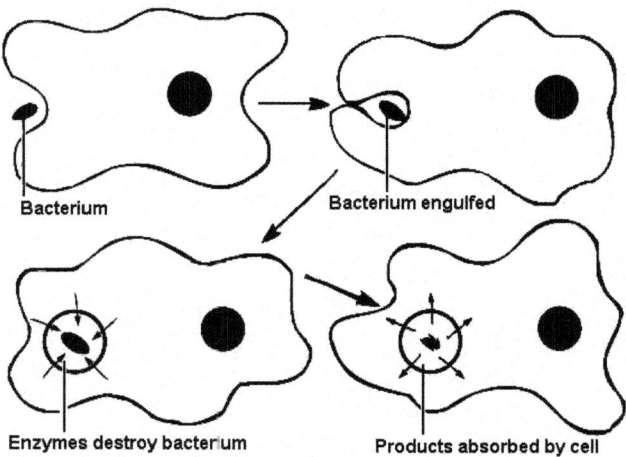

Bacterium   Bacterium engulfed

Enzymes destroy bacterium   Products absorbed by cell

Figure 9: Phagocytosis

Phagocytosis: Phagocytosis is sometimes called "cell eating". It is a process that requires energy and is used by cells to move solid particles like bacteria across the plasma membrane. Finger-like projections from the plasma membrane surround the bacteria and engulf them as shown. Once within the cell, enzymes produced by the lysosomes of the cell (described later) destroy the bacteria.

The destruction of bacteria and other foreign substance by white blood cells by the process of phagocytosis is a vital part of the defence mechanisms of the body.

Pinocytosis : Pinocytosis or "cell drinking" is a very similar process to phagocytosis but is used by cells to move fluids across the plasma membrane. Most cells carry out pinocytosis.

Exocytosis: Exocytosis is the process by means of which substances formed in the cell are moved through the plasma membrane into the fluid outside the cell (or extra-cellular fluid). It occurs in all cells but is most important in secretory cells (e.g. cells that produce digestive enzymes) and nerve cells.

The Cytoplasm

Within the plasma membrane is the cytoplasm. It consists of a clear jelly-like fluid called the a) cytosol or intracellular fluid in which b) cell inclusions, c) organelles and d) microfilaments and microtubules are found.

Cytosol : The cytosol consists mainly of water in which various molecules are dissolved or suspended. These molecules include proteins, fats and carbohydrates as well as sodium, potassium, calcium and chloride ions. Many of the reactions that take place in the cell occur in the cytosol.

Cell inclusions : These are large particles of fat, glycogen and melanin that have been produced by the cell. They are often large enough to be seen with the light microscope. For example the cells of adipose tissue (as in the insulating fat layer under the skin) contain fat that takes up most of the cell.

Organelles : Organelles are the "little organs" of the cell - like the heart, kidney and liver are the organs of the body. They are structures with characteristic appearances and specific "jobs" in the cell. Most can not be seen with the light microscope and so it was only when the electron microscope was developed that they were discovered. The main organelles in the cell are the ribosomes, endoplasmic reticulum, mitochondrion, Golgi complex and lysosomes.

Ribosomes

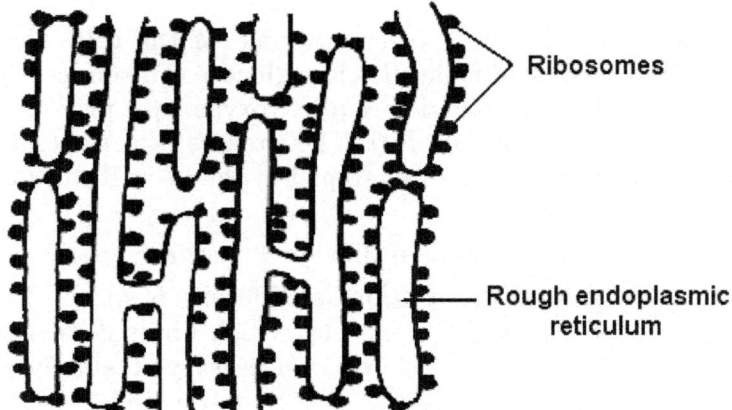

Figure 10: Rough endoplasmic reticulum

Ribosomes are tiny spherical organelles that make proteins by joining amino acids together. Many ribosomes are found free in the cytosol, while others are attached to the rough endoplasmic reticulum.

Endoplasmic Reticulum

The endoplasmic reticulum (ER) is a network of membranes that form channels throughout the cytoplasm from the nucleus to the plasma membrane. Various molecules are made in the ER and transported around the cell in its channels. There are two types of ER: smooth ER and rough ER.

Smooth ER is where the fats in the cell are made and in some cells, where chemicals like alcohol, pesticides and carcinogenic molecules are inactivated.

The Rough ER has ribosomes attached to its surface. The function of the Rough ER is therefore to make proteins that are modified stored and transported by the ER.

Mitochondria

Figure 11: A mitochondrion

Mitochondria (singular mitochondrion) are oval or rod shaped organelles scattered throughout the cytoplasm. They consist of two membranes, the inner one of which is folded to increase its surface area.

Mitochondria are the "power stations" of the cell. They make energy by "burning" food molecules like glucose. This process is called cellular respiration. The reaction requires oxygen and produces carbon dioxide which is a waste product. The process is very complex and takes place in a large number of steps but the overall word equation for cellular respiration is-

Glucose + oxygen = carbon dioxide + water + energy

Note that cellular respiration is different from respiration or breathing. Breathing is the means by which air is drawn into and expelled from the lungs. Breathing is necessary to supply the cells with the oxygen required by the mitochondria and to remove the carbon dioxide produced as a waste product of cellular respiration.

Active cells like muscle, liver, kidney and sperm cells have large numbers of mitochondria.

Golgi Apparatus

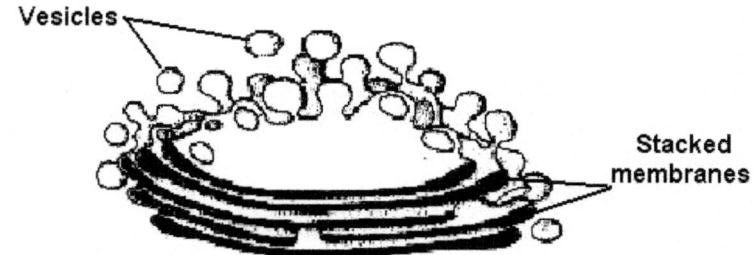

Figure 12: A Golgi body

The Golgi bodies in a cell together make up the Golgi apparatus. Golgi bodies are found near the nucleus and consist of flattened membranes stacked on top of each other rather like a pile of plates. The Golgi apparatus modifies and sorts the proteins and fats made by the ER, then surrounds them in a membrane as vesicles so they can be moved to other parts of the cell.

Lysosomes

Lysosomes are large vesicles that contain digestive enzymes. These break down bacteria and other substances that are brought into the cell by phagocytosis or pinocytosis. They also digest worn-out or damaged organelles, the components of which can then be recycled by the cell to make new structures.

Microfilaments And Microtubules

Some cells can move and change shape and organelles and chemicals are moved around the cell. Threadlike structures called microfilaments and microtubules that can contract are responsible for this movement. These structures also form the projections from the plasma membrane known as flagella (singular flagellum) as in the sperm tail, and cilia found lining the respiratory tract and used to remove mucus that has trapped dust particles.

Microtubules also form the pair of cylindrical structures called centrioles found near the nucleus. These help organise the spindle used in cell division.

The Nucleus

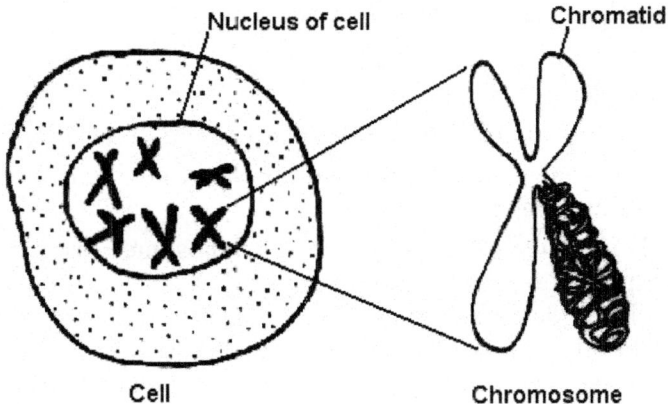

Figure 13: A cell with an enlarged chromosome

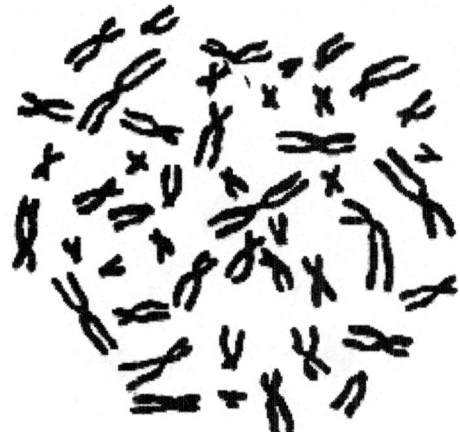

Figure 14: A full set of human chromosomes

The nucleus is the largest structure in a cell and can be seen with the light microscope. It is a spherical or oval body that contains the chromosomes. The nucleus controls the development and activity of the cell. Most cells contain a nucleus although mature red blood cells have lost theirs during development and some muscle cells have several nuclei. A double membrane similar in structure to the plasma membrane surrounds the nucleus (now called the nuclear envelope). Pores in this nuclear membrane allow communication between the nucleus and the cytoplasm. Within the nucleus one or more spherical bodies of darker material can be seen, even with the light microscope. These are called nucleoli and are made of RNA. Their role is to make new ribosomes.

Chromosomes

Inside the nucleus are the chromosomes which:

*   contain DNA;
*   control the activity of the cell;
*   are transmitted from cell to cell when cells divide;
*   are passed to a new individual when sex cells fuse together in sexual reproduction.

In cells that are not dividing the chromosomes are very long and thin and appear as dark grainy material. They become visible just before a cell divides when they shorten and thicken and can then be counted. The number of chromosomes in the cells of different species varies but is constant in the cells of any one species (e.g. horses have 64 chromosomes, cats have 38 and humans 46). Chromosomes occur in pairs (i.e. 32 pairs in the horse nucleus and 19 in that of the cat). Members of each pair are identical in length and shape and if you look carefully at diagram, you may be able to see some of the pairs in the human set of chromosomes.

Cell Division

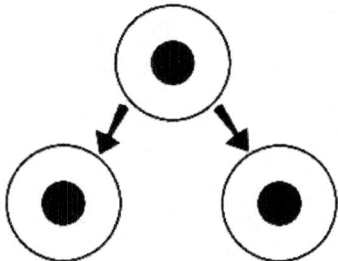

Figure 15: Division by mitosis results in 2 new cells identical to each other and to parent cell

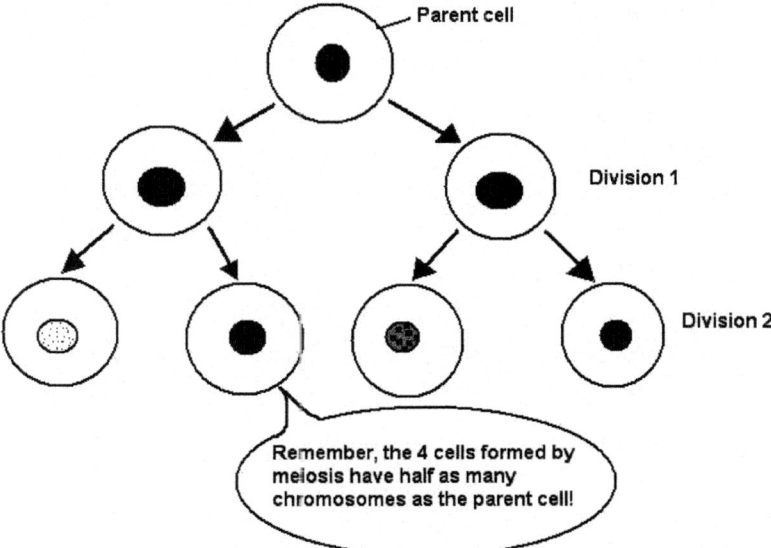

Parent cell

Division 1

Division 2

Remember, the 4 cells formed by meiosis have half as many chromosomes as the parent cell!

Figure 16: Division by meiosis results in 4 new cells that are genetically different to each other

Cells divide when an animal grows, when its body repairs an injury and when it produces sperm and eggs (or ova). There are two types of cell division: Mitosis and meiosis.

Mitosis: This is the cell division that occurs when an animal grows and when tissues are repaired or replaced. It produces two new cells (daughter cells) each with a full set of chromosomes that are identical to each other and to the parent cell. All the cells of an animal's body therefore contain identical DNA.

Meiosis: This is the cell division that produces the ova and sperm necessary for sexual reproduction. It only occurs in the ovary and testis.

The most important function of meiosis it to halve the number of chromosomes so that when the sperm fertilizes the ovum the normal number is regained. Body cells with the full set of chromosomes are called diploid cells, while gametes (sperm and ova) with half the chromosomes are called haploid cells.

Meiosis is a more complex process than mitosis as it involves two divisions one after the other and the four cells produced are all genetically different from each other and from the parent cell.

This fact that the cells formed by meiosis are all genetically different from each other and from the parent cell can be seen in litters of kittens where all the members of the litter are different from each other as well as being different from the parents although they display characteristics of both.

The Cell as a Factory

To make the function of the parts of the cell easier to understand and remember you can compare them to a factory. For example:

- The nucleus (1) is the managing director of the factory consulting the blueprint (the chromosomes) (2);
- The mitochondria (3) supply the power
- The ribosomes (4) make the products;
- The chloroplasts of plant cells (5) supply the fuel (food)
- The Golgi apparatus (6) packages the products ready for dispatch;
- The ER (7) modifies, stores and transports the products around the factory;
- The plasma membrane is the factory wall and the gates (8);
- The lysosomes dispose of the waste and worn-out machinery.

Body Organisation

The Organisation Of Animal Bodies

Living organisms move, feed, respire (burn food to make energy), grow, sense their environment, excrete and reproduce. These seven characteristics are sometimes summarised by the words "MRS GREN". functions of:

- Movement
- Respiration
- Sensitivity
- Growth
- Reproduction
- Excretion
- Nutrition.

Living organisms are made from cells which are organised into tissues and these are themselves combined to form organs and systems.

Skin cells, muscle cells, skeleton cells and nerve cells, for example. These different types of cells are not just scattered around randomly but similar cells that perform the same function are arranged in groups. These collections of similar cells are known as tissues.

There are four main types of tissues in animals. These are:

- Epithelial tissues that form linings, coverings and glands,
- Connective tissues for transport and support
- Muscle tissues for movement and
- Nervous tissues for carrying messages.

Epithelial Tissues

Epithelium (plural epithelia) is tissue that covers and lines. It covers an organ or lines a tube or space in the body. There are several different types of epithelium, distinguished by the different shapes of the cells and whether they consist of only a single layer of cells or several layers of cells.

Simple Epithelia - with a Single Layer of Cells

Squamous epithelium consists of a single layer of flattened cells that are shaped rather like 'crazy paving'. It is found lining the heart, blood vessels, lung alveoli and body cavities. Its thinness allows molecules to diffuse across readily.

Cuboidal Epithelium

Cuboidal epithelium consists of a single layer of cube shaped cells. It is rare in the body but is found lining kidney tubules. Molecules pass across it by diffusion, osmosis and active transport.

Columnar Epithelium

Columnar epithelium consists of column shaped cells. It is found lining the gut from the stomach to the anus. Digested food products move across it into the blood stream.

Columnar Epithelium with Cilia

Columnar epithelium with cilia on the free surface (also known as the apical side of the cell) lines the respiratory tract, fallopian tubes and uterus. The cilia beat rhythmically to transport particles.

Transitional Epithelium - with a Variable Number of Layers

The cells in transitional epithelium can move over one another allowing it to stretch. It is found in the wall of the bladder.

Stratified Epithelia - with Several Layers of Cells

Epithelia with several layers of cells are found where toughness and resistance to abrasion are needed.

Stratified Squamous Epithelium

Stratified squamous epithelium has many layers of flattened cells. It is found lining the mouth, cervix and vagina. Cells at the base divide and push up the cells above them and cells at the top are worn or pushed off the surface. This type of epithelium protects underlying layers and repairs itself rapidly if damaged.

Keratinised Stratified Squamous Epithelium

Keratinised stratified squamous epithelium has a tough waterproof protein called keratin deposited in the cells. It forms the skin found covering the outer surface of mammals.

Connective Tissues

Blood, bone, tendons, cartilage, fibrous connective tissue and fat (adipose) tissue are all classed as connective tissues. They are tissues that are used for supporting the body or transporting substances around the body. They also consist of three parts: they all have cells suspended in a ground substance or matrix and most have fibres running through it.

Blood

Blood consists of a matrix - plasma, with several types of cells and cell fragments suspended in it. The fibres are only evident in blood that has clotted. Blood will be described in detail in chapter 8.

Lymph

Lymph is similar in composition to blood plasma with various types of white blood cell floating in it. It flows in lymphatic vessels.

Connective Tissue 'Proper'

Connective tissue 'proper' consists of a jelly-like matrix with a dense network of collagen and elastic fibres and various cells embedded in it. There are various different forms of 'proper' connective tissue.

Loose Connective Tissue

Loose connective tissue is a sticky whitish substance that fills the spaces between organs. It is found in the dermis of the skin.

Dense Connective Tissue

Dense connective tissue contains lots of thick fibres and is very strong. It forms tendons, ligaments and heart valves and covers bones and organs like the kidney and liver.

Adipose Tissue

Adipose tissue consists of cells filled with fat. It forms the fatty layer under the dermis of the skin, around the kidneys and heart and the yellow marrow of the bones.

Cartilage

Cartilage is the 'gristle' of the meat. It consists of a tough jelly-like matrix with cells suspended in it. It may contain collagen and elastic fibres. It is a flexible but tough tissue and is found at the ends of bones, in the nose, ear and trachea and between the vertebrae .

Bone

Bone consists of a solid matrix made of calcium salts that give it its hardness. Collagen fibres running through it give it its strength. Bone cells are found in spaces in the matrix. Two types of bone are found in the skeleton namely spongy and compact bone. They differ in the way the cells and matrix are arranged.

Smooth Muscle

Smooth muscle consists of long and slender cells with a central nucleus. It is found in the walls of blood vessels, airways to the lungs and the gut. It changes the size of the blood vessels and helps move food and fluid along. Contraction of smooth muscle fibres occurs without the conscious control of the animal.

Skeletal Muscle

Skeletal muscle (sometimes called striated, striped or voluntary muscle) has striped fibres with alternating light and dark bands. It is attached to bones and is under the voluntary control of the animal.

Cardiac Muscle

Cardiac muscle is found only in the walls of the heart where it produces the 'heart beat'. Cardiac muscle cells are branched cylinders with central nuclei and faint stripes. Each fibre contracts automatically but the heart beat as a whole is controlled by the pacemaker and the involuntary autonomic nervous system.

Nervous Tissues

Nervous tissue forms the nerves, spinal cord and brain. Nerve cells or neurons consist of a cell body and a long thread or axon that carries the nerve impulse. An insulating sheath of fatty material (myelin) usually surrounds the axon. A typical motor neuron that sends messages to muscles to contract.

Vertebrate Bodies

We are so familiar with animals with backbones (i.e. vertebrates) that it seems rather unnecessary to point out that the body is divided into three sections. There is a well-defined head that contains the brain, the major sense organs and the mouth, a trunk that contains the other organs and a well-developed tail.

Other features of vertebrates may be less apparent. For instance, vertebrates that live on the land have developed a flexible neck that is absent in fish where it would be in the way of the gills and interfere with streamlining. Mammals but not other vertebrates have a sheet of muscle called the diaphragm that divides the trunk into the chest region or thorax and the abdomen.

Body Cavities

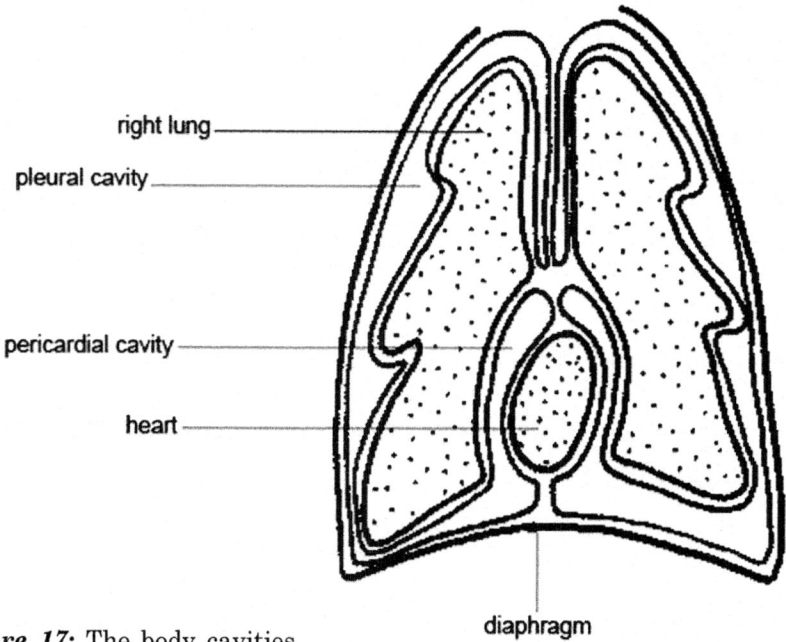

right lung

pleural cavity

pericardial cavity

heart

diaphragm

Figure 17: The body cavities

In contrast to many primitive animals, vertebrates have spaces or body cavities that contain the body organs. Most vertebrates have a single body cavity but in mammals the diaphragm divides the main cavity into a thoracic and an abdominal cavity. In the thoracic cavity the heart and lungs are surrounded by their own membranes so that cavities are created around the heart - the pericardial cavity, and around the lungs – the pleural cavity.

Organs

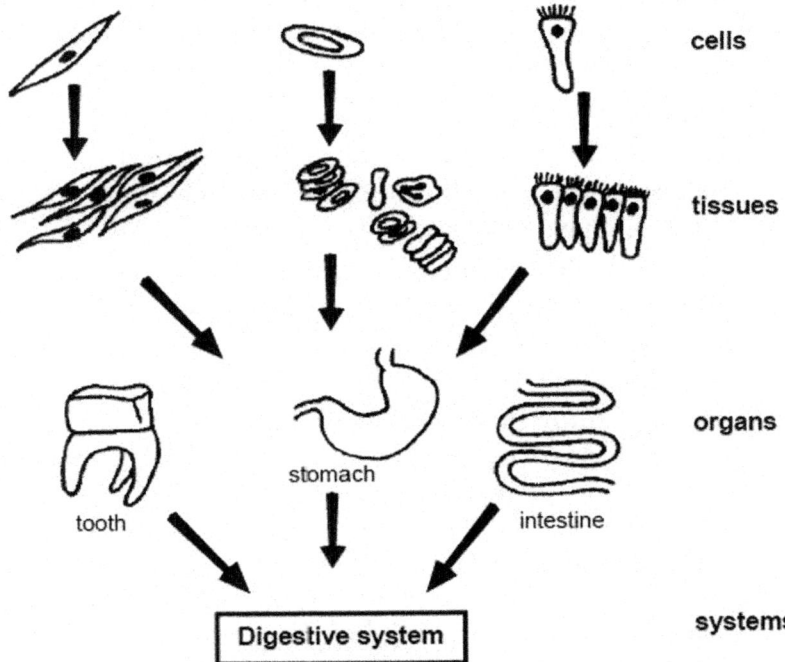

Figure 18: Cells, tissues and organs forming the digestive system

Just as the various parts of the cell work together to perform the cell's functions and a large number of similar cells make up a tissue, so many different tissues can "cooperate" to form an organ that performs a particular function. For example, connective tissues, epithelial tissues, muscle tissue and nervous tissue combine to make the organ that we call the stomach. In turn the stomach combines with other organs like the intestines, liver and pancreas to form the digestive system.

Generalised Plan of The Mammalian Body

At this point it would be a good idea to make yourself familiar with the major organs and their positions in the body of a mammal like the rabbit.

Body Systems

Organs do not work in isolation but function in cooperation with other organs and body structures to bring about the MRS GREN functions necessary to keep an animal alive. For example the stomach can only work in conjunction with the mouth and oesophagus (gullet).

These provide it with the food it breaks down and digests. It then needs to pass the food on to the intestines etc. for further digestion and absorption. The organs involved with the taking of food into the body, the digestion and absorption of the food and elimination of waste products are collectively known as the digestive system.

The 11 body Systems

1. *Skin:* The skin covering the body consists of two layers, the epidermis and dermis. Associated with these layers are hairs, feathers, claws, hoofs, glands and sense organs of the skin.

2. *Skeletal System:* This can be divided into the bones of the skeleton and the joints where the bones move over each other.

3. *Muscular System:* The muscles, in conjunction with the skeleton and joints, give the body the ability to move.

4. *Cardiovascular System:* This is also known as the circulatory system. It consists of the heart, the blood vessels and the blood. It transports substances around the body.

5. *Lymphatic System:* This system is responsible for collecting and "cleaning" the fluid that leaks out of the blood vessels. This fluid is then returned to the blood system. The lymphatic system also makes antibodies that protect the body from invasion by bacteria etc. It consists of lymphatic vessels, lymph nodes, the spleen and thymus glands.

6. *Respiratory System:* This is the system involved with bringing oxygen in the air into the body and getting rid of carbon dioxide, which is a waste product of processes that occur in the cell. It is made up of the trachea, bronchi, bronchioles, lungs, diaphragm, ribs and muscles that move the ribs in breathing.

7. *Digestive System:* This is also known as the gastrointestinal system, alimentary system or gut. It consists of the digestive tube and glands like the liver and pancreas that produce digestive secretions. It is concerned with breaking down the large molecules in foods into smaller ones that can be absorbed into the blood and lymph. Waste material is also eliminated by the digestive system.

8. *Urinary System:* This is also known as the renal system. It removes waste products from the blood and is made up of the kidneys, ureters and bladder.

9. *Reproductive System:* This is the system that keeps the species going by making new individuals. It is made up of the ovaries,

uterus, vagina and fallopian tubes in the female and the testes with associated glands and ducts in the male.

10. *Nervous System:* This system coordinates the activities of the body and responses to the environment. It consists of the sense organs (eye, ear, semicircular canals, and organs of taste and smell), the nerves, brain and spinal cord.

11. Endocrine System : This is the system that produces chemical messengers or hormones. It consists of various endocrine glands (ductless glands) that include the pituitary, adrenal, thyroid and pineal glands as well as the testes and ovary.

Homeostasis

All the body systems, except the reproductive system, are involved with keeping the conditions inside the animal more or less stable. This is called homeostasis. These constant conditions are essential for the survival and proper functioning of the cells, tissues and organs of the body. The skin, for example, has an important role in keeping the temperature of the body constant. The kidneys keep the concentration of salts in the blood within limits and the islets of Langerhans in the pancreas maintain the correct level of glucose in the blood through the hormone insulin. As long as the various body processes remain within normal limits, the body functions properly and is healthy. Once homeostasis is disturbed disease or death may result.

Directional Terms

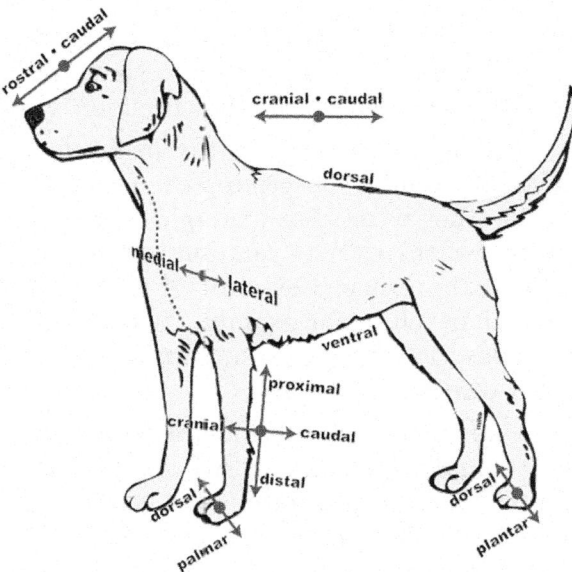

Figure 19: The directional terms used with animals

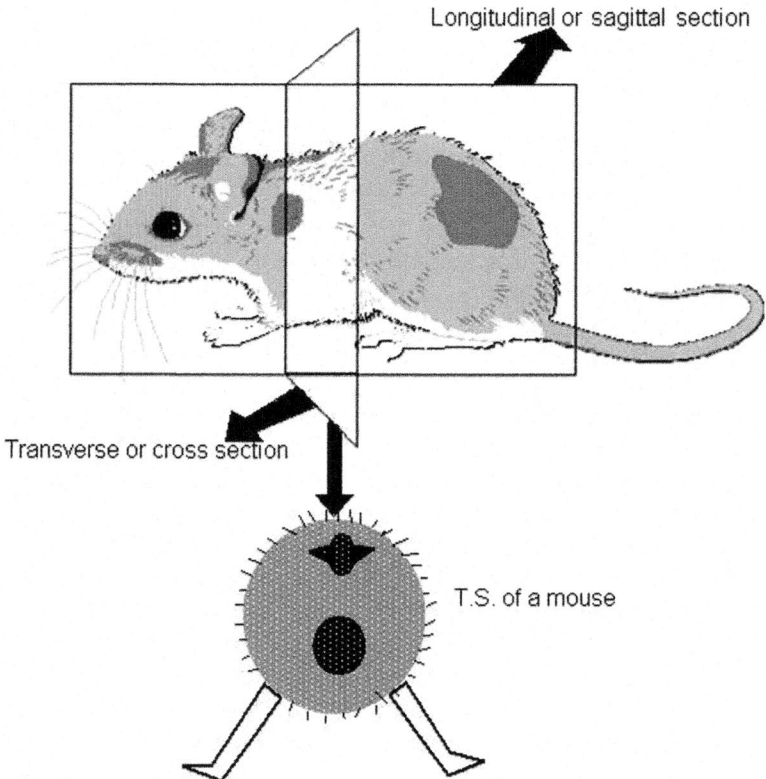

Figure 20: Transverse and longitudinal sections of a mouse

In the following chapters the systems of the body in the list above will be covered one by one. For each one the structure of the organs involved will be described and the way they function will be explained.

In order to describe structures in the body of an animal it is necessary to have a system for describing the position of parts of the body in relation to other parts. For example it may be necessary to describe the position of the liver in relation to the diaphragm, or the heart in relation to the lungs. Certainly if you work further with animals, in a veterinary clinic for example, it will be necessary to be able to accurately describe the position of an injury. The terms used for this are called directional terms.

The most common directional terms are right and left. However, even these are not completely straightforward especially when looking at diagrams of animals. The convention is to show the left side of the animal or organ on the right side of the page. This is the view you would get looking down on an animal lying on its back during surgery or in a post-mortem. Sometimes it is useful to imagine 'getting inside' the animal (so to speak) to check which side is which. The other common and useful directional terms are listed below and shown.

In order to look at the structure of some of the parts or organs of the body it may be necessary to cut them open or even make thin slices of them that they can be examined under the microscope. The direction and position of slices or sections through an animal's body have their own terminology. If an animal or organ is sliced lengthwise this section is called a longitudinal or sagittal section. This is sometimes abbreviated to LS.

If the section is sliced crosswise it is called a transverse or cross section. This is sometimes abbreviated to TS or XS.

The Skin

Skin comes in all kinds of textures and forms. There is the dry warty skin of toads and crocodiles, the wet slimy skin of fish and frogs, the hard shell of tortoises and the soft supple skin of snakes and humans. Mammalian skin is covered with hair, that of birds with feathers, and fish and reptiles have scales. Pigment in the skin, hairs or feathers can make the outer surface almost any colour of the rainbow.

As humans, it is often the skin of an animal that gives it its appeal to us or repels us. We love the soft feel of a cat's coat but perhaps can't bear to touch a snake. As the main part of an animal visible to us, the skin can often give us clues to the health of an animal. A healthy animal will have a clean, glowing, flexible skin, while ill health may show itself as an abnormal colour or texture.

Skin is one of the largest organs of the body, making up 6-8% of the total body weight. It consists of two distinct layers. The top layer is called the epidermis and under that is the dermis.

The epidermis is the layer that bubbles up when we have a blister and as we know from this experience, it has no blood or nerves in it. The cells at the base of the epidermis continually divide and push the cells above them upwards. As these cells move up they die and become the dry flaky scales that fall off the skin surface. The cells in the epidermis die because a special protein called keratin is deposited in them. Keratin is an extremely important substance for it makes the skin waterproof. Without it land vertebrates like reptiles, birds and mammals would, like frogs, be able to survive only in damp places.

Skin Structures Made of Keratin

Claws, Nails and Hoofs: Reptiles, birds and mammals all have nails or claws on the ends of their toes. They protect the end of the toe and may be used for grasping, grooming, digging or in defence. They are continually worn away and grow continuously from a growth layer at their base.

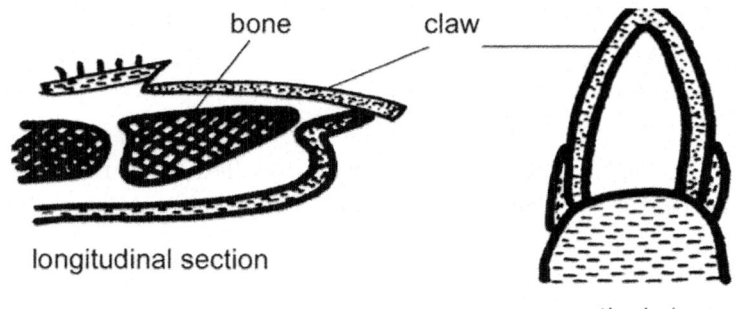

longitudinal section

vertical view

Figure 21: A carnivore's claw

Hoofs are found in sheep, cows, horses etc. otherwise known as ungulate mammals. These are animals that have lost toes in the process of evolution and walk on the "nails" of the remaining toes. The hoof is a cylinder of horny material that surrounds and protects the tip of the toe.

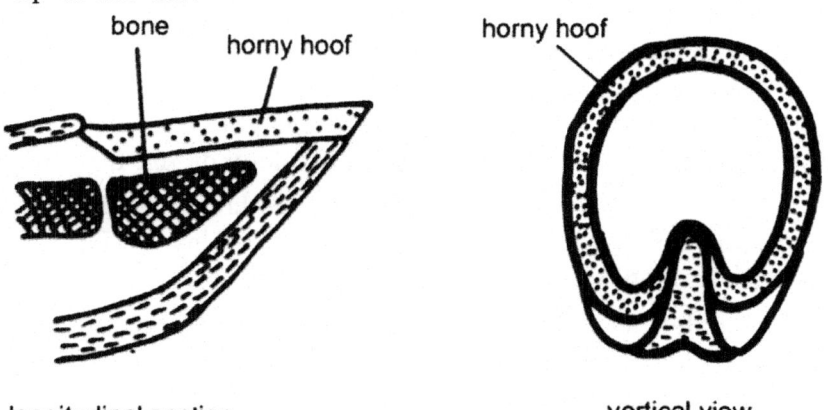

longitudinal section   vertical view

Figure 22: A horse's hoof

Horns And Antlers

True horns are made of keratin and are found in sheep, goats and cattle. They are never branched and, once grown, are never shed. They consist of a core of bone arising in the dermis of the skin and are fused with the skull. The horn itself forms as a hollow cone-shaped sheath around the bone.

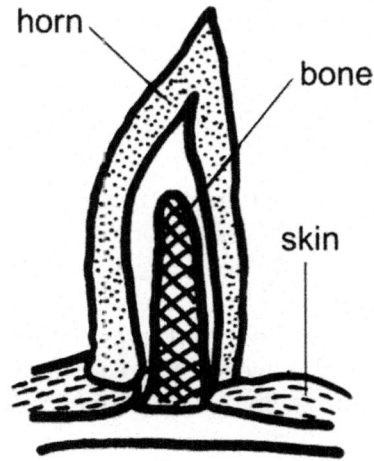

Figure 23: A horn

The antlers of male deer have quite a different structure. They are not formed in the epidermis and do not consist of keratin but are entirely of bone. They are shed each year and are often branched, especially in older animals. When growing they are covered in skin called velvet that forms the bone. Later the velvet is shed to leave the bony antler. The velvet is often removed artificially to be sold in Asia as a traditional medicine.

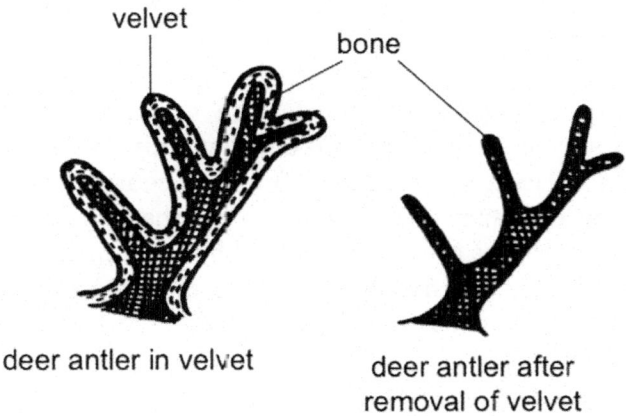

deer antler in velvet   deer antler after
                                   removal of velvet

Figure 24: A deer antler

Other animals have projections on their heads that are not true horns either. The horns on the head of giraffes are made of bone covered with skin and hair, and the 'horn' of a rhinoceros is made of modified and fused hair-like structures.

Hair

Hair is also made of keratin and develops in the epidermis. It covers the body of most mammals where it acts as an insulator and helps to regulate the temperature of the body. The colour in hairs is formed from the same pigment, melanin that colours the skin. Coat colour may help camouflage animals and sometimes acts to attract the opposite sex.

Hairs lie in a follicle and grow from a root that is well supplied with blood vessels. The hair itself consists of layers of dead keratin - containing cells and usually lies at a slant in the skin. A small bundle of smooth muscle fibres (the hair erector muscle) is attached to the side of each hair and when this contracts the hair stands on end. This increases the insulating power of the coat and is also used by some animals to make them seem larger when confronted by a foe or a competitor.

The whiskers of cats and the spines of hedgehogs are examples of special types of hairs.

Feathers

The lightness and stiffness of keratin is also a key to bird flight. In the form of feathers it provides the large airfoils necessary for flapping and gliding flight. In another form, the light fluffy down feathers,also made of keratin, are some of the best natural insulators known. This superior insulation is necessary to help maintain the high body temperatures of birds.

Countour feathers are large feathers that cover the body, wings and tail. They have an expanded vane that provides the smooth, continuous surface that is required for effective flight. This surface is formed by barbs that extend out from the central shaft. If you look carefully at a feather you can see that on either side of each barb are thousands of barbules that lock together by a complex system of hooks and notches. If this arrangement becomes disrupted, the bird uses its beak to draw the barbs and barbules together again in an action known as preening.

Down feathers are the only feathers covering a chick and form the main insulation layer under the contour feathers of the adult. They have no shaft but consist of a spray of simple, slender branches.

Pin feathers have a slender hair-like shaft often with a tiny tuft of barbs on the end. They are found between the other feathers and help tell a bird how its feathers are lying.

Skin Glands

Glands are organs that produce and secrete fluids. They are usually divided into two groups depending upon whether or not they have channels or ducts to carry their products away. Glands with ducts are called exocrine glands and include the glands found in the skin as well as the glands that produce digestive enzymes in the gut. Endocrine glands have no ducts and release their products (hormones) directly into the blood stream. The pituitary and adrenal glands are examples of endocrine glands.

Most vertebrates have exocrine glands in the skin that produce a variety of secretions. The slime on the skin of fish and frogs is mucus produced by skin glands and some fish and frogs also produce poison from modified glands.

In fact the skin glands of some frogs produce the most poisonous chemicals known. Reptiles and birds have a dry skin with few glands. The preen gland, situated near the base of the bird's tail, produces oil to help keep the feathers in good condition. Mammals have an array of different skin glands. These include the wax producing, sweat, sebaceous and mammary glands.

Wax producing glands are found in the ears.

Sebaceous glands secrete an oily secretion into the hair follicle. This secretion, known as sebum, keeps the hair supple and helps prevent the growth of bacteria.

Sweat glands consist of a coiled tube and a duct leading onto the skin surface. Their appearance when examined under the microscope inspired one of the first scientists to observe them to call them "fairies' intestines".

Sweat contains salt and waste products like urea and the evaporation of sweat on the skin surface is one of the major mechanisms for cooling the body of many mammals. Horses can sweat up to 30 litres of fluid a day during active exercise, but cats and dogs have few sweat glands and must cool themselves by panting. Scent in the sweat of many animals is used to mark territory or attract the opposite sex.

Mammary glands are only present in mammals. They are thought to be modified sebaceous glands and are present in both sexes but are rarely active in males. The number of glands varies from species to species. They open to the surface in well-developed nipples. Milk contains proteins, sugars, fats and salts, although the exact composition varies from one species to another.

The Skin And Sun

A moderate amount of UV in sunlight is necessary for the skin to form vitamin D. This vitamin prevents bone disorders like rickets to which animals reared indoors are susceptible. Excessive exposure to the UV in sunlight can be damaging and the pigment melanin, deposited in cells at the base of the epidermis, helps to protect the underlying layers of the skin from this damage. Melanin also colours the skin and variations in the amount of melanin produces colours from pale yellow to black.

Sunburn And Skin Cancer

Excess exposure to the sun can cause sunburn. This is common in humans, but light skinned animals like cats and pigs can also be sunburned, especially on the ears. Skin cancer can also result from excessive exposure to the sun. As holes in the ozone layer increase exposure to the sun's UV rays, so too does the rate of skin cancer in humans and animals.

The Dermis

The underlying layer of the skin, known as the dermis, is much thicker but much more uniform in structure than the epidermis. It is composed of loose connective tissue with a felted mass of collagen and elastic fibres. It is this part of the skin of cattle and pigs etc. that becomes commercial leather when treated,. The dermis is well supplied with blood vessels, so cuts and burns that penetrate down into the dermis will bleed or cause serious fluid loss. There are also numerous nerve endings and touch receptors in the dermis because, of course, the skin is sensitive to touch, pain and temperature.

When looking at a section of the skin under the microscope you can see hair follicles, sweat and sebaceous glands dipping down into the dermis. However, these structures do not originate in the dermis but are derived from the epidermis.

In the lower levels of the dermis is a layer of fat or adipose tissue. This acts as an energy store and is an excellent insulator especially in mammals like whales with little hair.

The Skin And Temperature Regulation

Vertebrates can be divided into two groups depending on whether or not they control their internal temperature. Amphibia (frogs) and reptiles are said to be cold blooded" (poikilothermic) because their body temperature approximately follows that of the environment. Birds and mammals are said to be warm blooded (homoiothermic) because they can maintain a roughly constant body temperature despite changes in the temperature of the environment.

Heat is produced by the biochemical reactions of the body (especially in the liver) and by muscle contraction. Most of the heat lost from the body occurs via the skin. It is therefore not surprising that many of the mechanisms for controlling the temperature of the body operate here.

Reduction of Heat Loss

When an animal is in a cold environment and needs to reduce heat loss the erector muscles contract causing the hair or feathers to rise up and increase the layer of insulating air trapped by them.

Heat loss from the skin surface can also be reduced by the contraction of the abundant blood vessels that lie in the dermis. This takes blood flow to deeper levels, so reducing heat loss and causing pale skin.

Shivering caused by twitching muscles produces heat that also helps raise the body temperature.

Increase of Heat Loss

There are two main mechanisms used by animals to increase the amount of heat lost from the skin when they are in a hot environment or high levels of activity are increasing internal heat production. The first is the expansion of the blood vessels in the dermis so blood flows near the skin surface and heat loss to the environment can take place. The second is by the production of sweat from the sweat glands. The evaporation of this liquid on the skin surface produces a cooling effect.

The mechanisms for regulating body temperature are under the control of a small region of the brain called the hypothalamus. This acts like a thermostat.

Heat Loss and Body Size

The amount of heat that can be lost from the surface of the body is related to the area of skin an animal has in relation to the total volume of its body.

Small animals like mice have a very large skin area compared to their total volume. This means they tend to loose large amounts of heat and have difficulty keeping warm in cold weather. They may need to keep active just to maintain their body temperature or may hibernate to avoid the problem.

Large animals like elephants have the opposite problem. They have only a relatively small skin area in relation to their total volume and may have trouble keeping cool. This is one reason that these large animals tend to have sparse coverings of hair.

Anatomy and Physiology of Animals/The Skeleton

Fish, frogs, reptiles, birds and mammals are called vertebrates, a name that comes from the bony column of vertebrae (the spine) that supports the body and head. The rest of the skeleton of all these animals (except the fish) also has the same basic design with a skull that houses and protects the brain and sense organs and ribs that protect the heart and lungs and, in mammals, make breathing possible. Each of the four limbs is made to the same basic pattern. It is joined to the spine by means of a flat, broad bone called a girdle and consists of one long upper bone, two long lower bones, several smaller bones in the wrist or ankle and five digits.

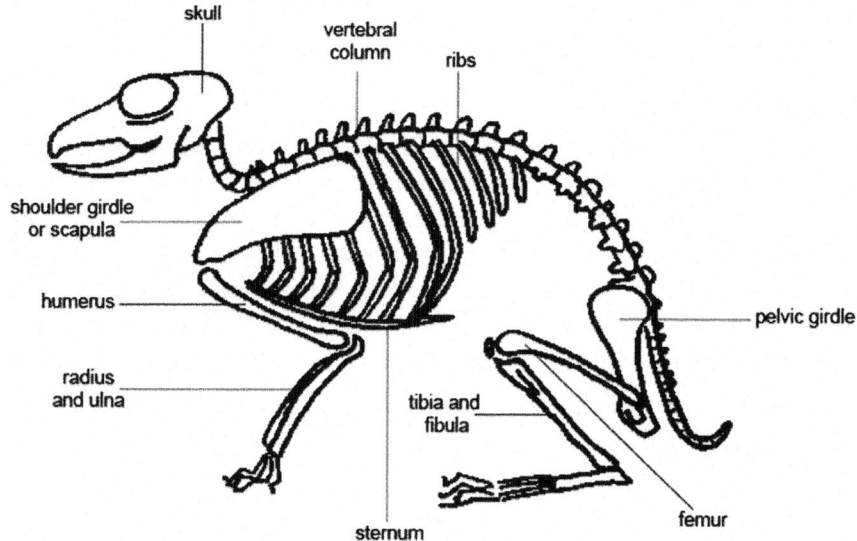

Figure 25: The mammalian skeleton

The Vertebral Column

The vertebral column consists of a series of bones called vertebrae linked together to form a flexible column with the skull at one end and the tail at the other. Each vertebra consists of a ring of bone with spines (spinous process) protruding dorsally from it. The spinal cord passes through the hole in the middle and muscles attach to the spines making movement of the body possible.

The shape and size of the vertebrae of mammals vary from the neck to the tail. In the neck there are cervical vertebrae with the two top ones, the atlas and axis, being specialised to support the head and allow it to nod "Yes" and shake "No". Thoracic vertebrae in the chest region have special surfaces against which the ribs move during breathing. Grazing animals like cows and giraffes that have to support weighty heads on long necks have extra large spines on their cervical and thoracic vertebrae for muscles to attach to. Lumbar vertebrae in the loin region are usually large strong vertebrae with prominent spines for the attachment of the large muscles of the lower back. The sacral vertebrae are usually fused into one solid bone called the sacrum that sits within the pelvic girdle. Finally there are a variable number of small bones in the tail called the coccygeal vertebrae.

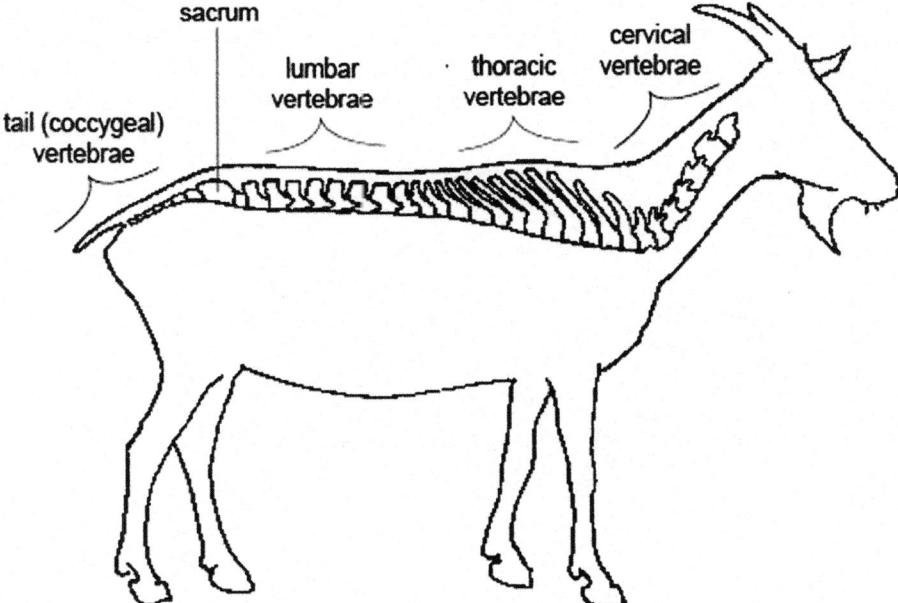

Figure 26: The regions of the vertebral column dik

The Skull

The skull of mammals consists of 30 separate bones that grow together during development to form a solid case protecting the brain and sense organs. The "box "enclosing and protecting the brain is called the cranium. The bony wall of the cranium encloses the middle and inner ears, protects the organs of smell in the nasal cavity and the eyes in sockets known as orbits. The teeth are inserted into the upper and lower jaws. The lower jaw is known as the mandible. It forms a joint with the skull moved by strong muscles that allow an animal to chew. At the front of the skull is the nasal cavity, separated from the mouth by a plate of bone called the palate. Behind the nasal cavity and connecting with it are the sinuses. These are air spaces in the bones of the skull which help keep the skull as light as possible. At the base of the cranium is the foramen magnum, translated as "big hole", through which the spinal cord passes. On either side of this are two small, smooth rounded knobs or condyles that articulate (move against) the first or Atlas vertebra.

The Rib

Paired ribs are attached to each thoracic vertebra against which they move in breathing. Each rib is attached ventrally either to the sternum or to the rib in front by cartilage to form the rib cage that protects the heart and lungs. In dogs one pair of ribs is not attached ventrally at all. They are called floating ribs. Birds have a large expanded sternum called the keel to which the flight muscles (the 'breast" meat of a roast chicken) are attached.

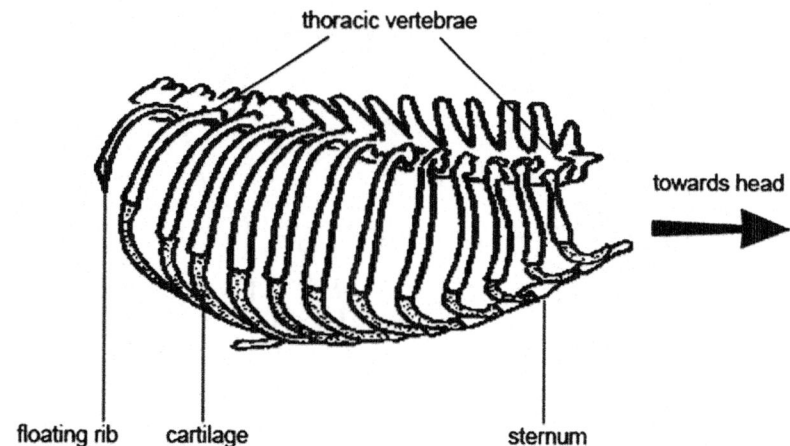

Figure 27: The rib

The Forelimb

The forelimb consists of: Humerus, radius and ulna, carpals, metacarpals, digits or phalanges. The top of the humerus moves against (articulates with) the scapula at the shoulder joint. By changing the number, size and shape of the various bones, fore limbs have evolved to fit different ways of life. They have become wings for flying in birds and bats, flippers for swimming in whales, seals and porpoises, fast and efficient limbs for running in horses and arms and hands for holding and manipulating in primates.

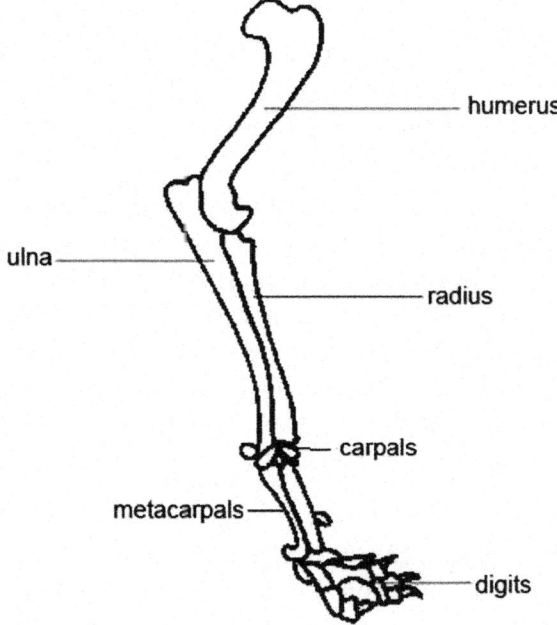

Figure 28: Forelimb of a dog

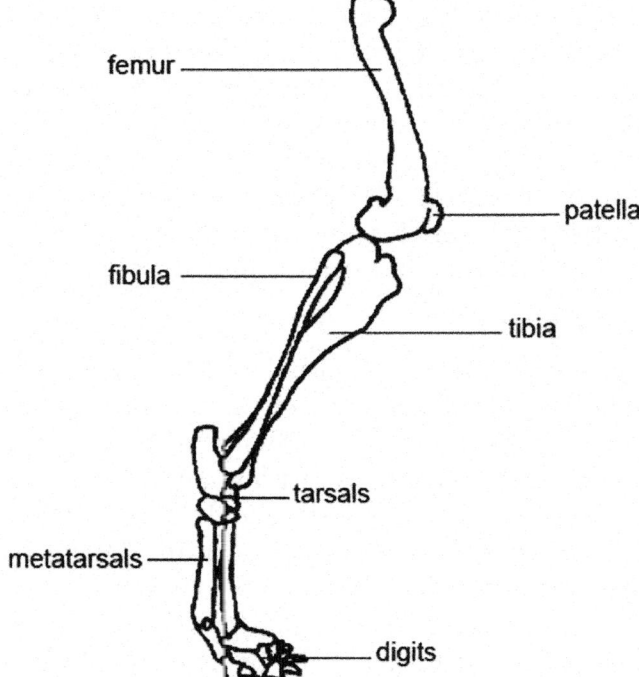

Figure 29: Hindlimb of a dog

The Hind Limb

The hind limbs have a similar basic pattern to the forelimb. They consist of: femur, tibia and fibula, tarsals, metatarsals, digits or phalanges. The top of the femur moves against (articulates with) the pelvis at the hip joint.

The Long pastern or proximal phalanx is P1, not P3. The distal phalanx or coffin bone (called hoof there) is the 3rd phalanges. The patella or kneecap is embedded in a large tendon in front of the knee. It seems to smooth the movements of the knee. The legs of the horse are highly adapted to give it great galloping speed over long distances. The bones of the leg, wrist and foot are greatly elongated and the hooves are actually the tips of the third fingers and toes, the other digits having been lost or reduced.

The Girdles

The girdles pass on the "push" produced by the limbs to the body. The shoulder girdle or scapula is a triangle of bone surrounded by the muscles of the back but not connected directly to the spine. This arrangement helps it to cushion the body when landing after a leap and gives the forelimbs the flexibility to manipulate food or strike at prey. Animals that use their forelimbs for grasping, burrowing or climbing have a well-developed clavicle or collar bone. This connects the shoulder girdle to the sternum. Animals like sheep, horses and cows that use their forelimbs only for supporting the body and locomotion have no clavicle. The pelvic girdle or hipbone attaches the sacrum and the hind legs. It transmits the force of the leg-thrust in walking or jumping directly to the spine.

Categories of Bones

People who study skeletons place the different bones of the skeleton into groups according to their shape or the way in which they develop. Thus we have long bones like the femur, radius and finger bones, short bones like the ones of the wrist and ankle, irregular bones like the vertebrae and flat bones like the shoulder blade and bones of the skull. Finally there are bones that develop in tissue separated from the main skeleton. These include sesamoid bones which include bones like the patella or kneecap that develop in tendons and visceral bones that develop in the soft tissue of the penis of the dog and the cow's heart.

Bird Skeletons

Although the skeleton of birds is made up of the same bones as that of mammals, many are highly adapted for flight. The most noticeable difference is that the bones of the forelimbs are elongated to act as wings. The large flight muscles make up as much as 1/5th of the body weight and are attached to an extension of the sternum called the keel. The vertebrae of the lower back are fused to provide the rigidity needed to produce flying movements. There are also many adaptations to reduce the weight of the skeleton. For instance birds have a beak rather than teeth and many of the bones are hollow.

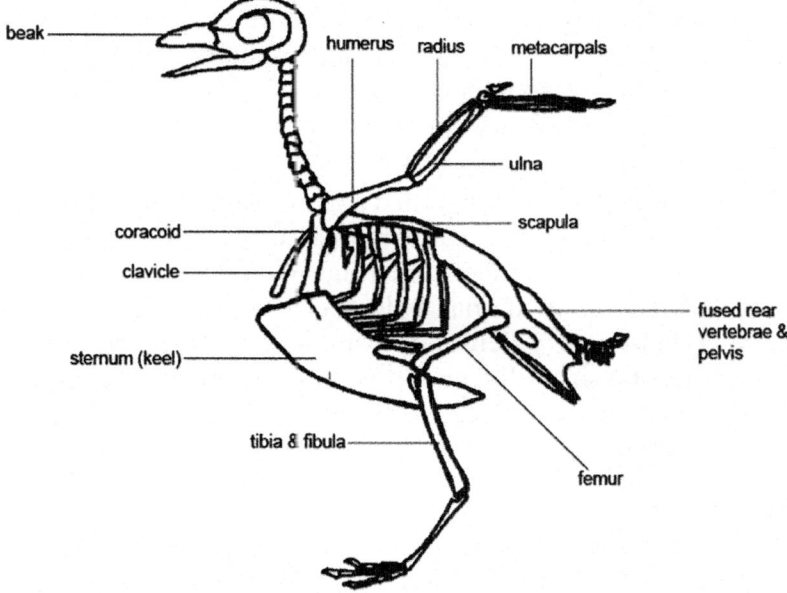

Figure 30: A bird's skeleton

The Structure of Long Bones

A long bone consists of a central portion or shaft and two ends called epiphyses. Long bones move against or articulate with other bones at joints and their ends have flattened surfaces and rounded protuberances (condyles) to make this possible. If you carefully examine a long bone you may also see raised or rough surfaces. This is where the muscles that move the bones are attached. You will also see holes (a hole is called a foramen) in the bone. Blood vessels and nerves pass into the bone through these. You may also be able to see a fine line at each end of the bone. This is called the growth plate or epiphyseal line and marks the place where increase in length of the bone occurred.

If you cut a long bone lengthways you will see it consists of a hollow cylinder. The outer shell is covered by a tough fibrous sheath to which the tendons are attached. Under this is a layer of hard, dense compact bone.

This gives the bone its strength. The central cavity contains fatty yellow marrow, an important energy store for the body, and the ends are made from honeycomb-like bony material called spongy bone. Spongy bone contains red marrow where red blood cells are made.

Compact Bone

Compact bone is not the lifeless material it may appear at first glance. It is a living dynamic tissue with blood vessels, nerves and living cells that continually rebuild and reshape the bone structure as a result of the stresses, bends and breaks it experiences. Compact bone is composed of microscopic hollow cylinders that run parallel to each other along the length of the bone.

Each of these cylinders is called a Haversian system. Blood vessels and nerves run along the central canal of each Haversian system. Each system consists of concentric rings of bone material (the matrix) with minute spaces in it that hold the bone cells. The hard matrix contains crystals of calcium phosphate, calcium carbonate and magnesium salts with collagen fibres that make the bone stronger and somewhat flexible. Tiny canals connect the cells with each other and their blood supply.

Spongy Bone

Spongy bone gives bones lightness with strength. It consists of an irregular lattice that looks just like an old fashioned loofah sponge. It is found on the ends of long bones and makes up most of the bone tissue of the limb girdles, ribs, sternum, vertebrae and skull. The spaces contain red marrow, which is where red blood cells are made and stored.

Bone Growth

The skeleton starts off in the foetus as either cartilage or fibrous connective tissue. Before birth and, sometimes for years after it, the cartilage is gradually replaced by bone. The long bones increase in length at the ends at an area known as the epiphyseal plate where new cartilage is laid down and then gradually converted to bone. When an animal is mature, bone growth ceases and the epiphyseal plate converts into a fine epiphyseal line.

Broken Bones

A fracture or break dramatically demonstrates the dynamic nature of bone. Soon after the break occurs blood pours into the site and cartilage is deposited. This starts to connect the broken ends together. Later spongy bone replaces the cartilage, which is itself replaced by compact bone. Partial healing to the point where some weight can be put on the bone can take place in 6 weeks but complete healing may take 3-4 months.

Joints

Joints are the structures in the skeleton where 2 or more bones meet. There are several different types of joints. Some are immovable once the animal has reached maturity. Examples of these are those between the bones of the skull and the midline joint of the pelvic girdle. Some are slightly moveable like the joints between the vertebrae but most joints allow free movement and have a typical structure with a fluid filled cavity separating the articulating surfaces (surfaces that move against each other) of the two bones. This kind of joint is called a synovial joint. The joint is held together by bundles of white fibrous tissue called ligaments and a fibrous capsule encloses the joint. The inner layers of this capsule secrete the synovial fluid that acts as a lubricant. The articulating surfaces of the bones are covered with cartilage that also reduces friction and some joints, e.g. the knee, have a pad of cartilage between the surfaces that articulate with each other.

The shape of the articulating bones in a joint and the arrangement of ligaments determine the kind of movement made by the joint. Some joints only allow a to and from gliding movement e.g. between the ankle and wrist bones; the joints at the elbow, knee and fingers are hinge joints and allow movement in two dimensions and the axis vertebra pivots on the atlas vertebra. Ball and socket joints, like those at the shoulder and hip, allow the greatest range of movement.

Common Names of Joints

Some joints in animals are given common names that tend to be confusing. For example:

1. The joint between the femur and the tibia on the hind leg is our knee but the stifle in animals.

2. Our ankle joint (between the tarsals and metatarsals) is the hock in animals

3. Our knuckle joint (between the metacarpals or metatarsals and the phalanges) is the fetlock in the horse.

4. The "knee" on the horse is equivalent to our wrist (ie on the front limb between the radius and metacarpals).

Locomotion

Different animals place different parts of the foot or forelimb on the ground when walking or running.

Humans and bears put the whole surface of the foot on the ground when they walk. This is known as plantigrade locomotion. Dogs and cats walk on their toes (digitigrade locomotion) while horses and pigs walk on their "toenails" or hoofs. This is called unguligrade locomotion.

1. Plantigrade locomotion (on the "palms of the hand) as in humans and bears

2. Digitigrade locomotion (on the "fingers") as in cats and dogs

3. Unguligrade locomotion (on the "fingernails") as in horses.

Muscles

Muscles make up the bulk of an animal's body and account for about half its weight. The meat on the chop or roast is muscle and is composed mainly of protein. The cells that make up muscle tissue are elongated and able to contract to a half or even a third of their length when at rest. There are three different kinds of muscle; smooth, cardiac and skeletal muscle.

Smooth Muscle

Smooth or Involuntary muscle carries out the unconscious routine tasks of the body such as moving food down the digestive system, keeping the eyes in focus and adjusting the diametre of blood vessels. The individual cells are spindle-shaped, being fatter in the middle and tapering off towards the ends with a nucleus in the centre of the cell. They are usually found in sheets and are stimulated by the non-conscious or autonomic nervous system as well as by hormones.

Cardiac Muscle

Cardiac muscle is only found in the wall of the heart. It is composed of branching fibres that form a three-dimensional network. When examined under the microscope, a central nucleus and faint stripes or striations can be seen in the cells. Cardiac muscle cells contract spontaneously and rhythmically without outside stimulation but the pacemaker coordinates the heart beat. Nerves and hormones modify this rhythm.

Skeletal Muscle

Skeletal muscle is the muscle that is attached to and moves the skeleton, and is under voluntary control. It is composed of elongated cells or fibres lying parallel to each other. Each cell is unusual in that it has several nuclei and when examined under the microscope appears striped or striated. This appearance gives the muscle its names of striped or striated muscle. Each cell of striated muscle contains hundreds, or even thousands, of microscopic fibres each one with its own striped appearance. The stripes are formed by two different sorts of protein that slide over each other making the cell contract.

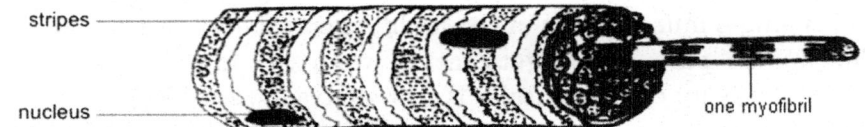

Figure 31: A striped muscle cell

Muscle Contraction

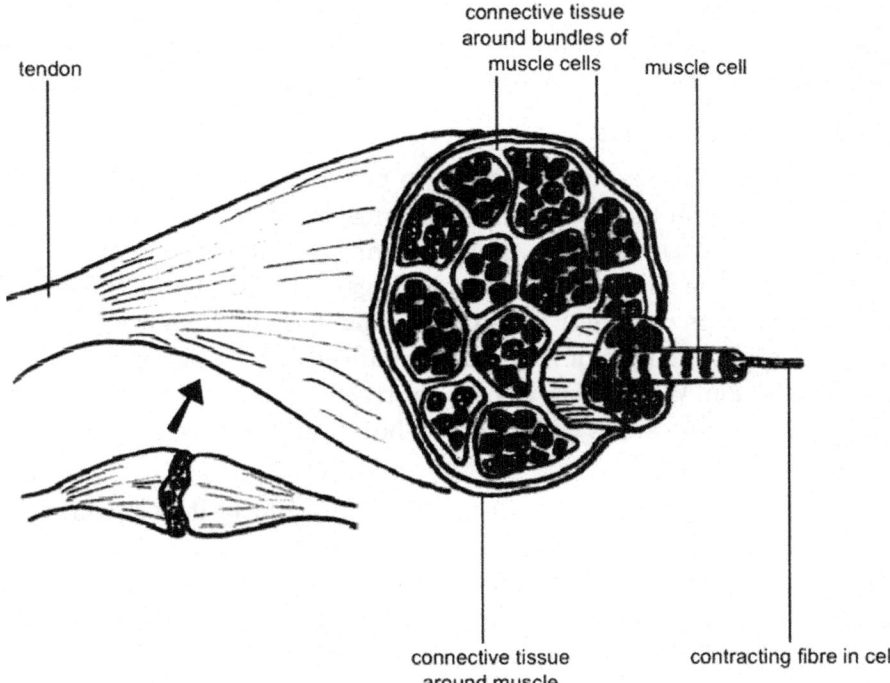

Figure 32: The structure of a muscle

Muscle contraction requires energy and muscle cells have numerous mitochondria. However, only about 15% of the energy released by the mitochondria is used to fuel muscle contraction. The rest is released as heat. This is why exercise increases body temperature and makes animals sweat or pant to rid themselves of this heat.

What we refer to as a muscle is made up of groups of muscle fibres surrounded by connective tissue. The connective tissue sheaths join together at the ends of the muscle to form tough white bands of fibre called tendons. These attach the muscles to the bones. Tendons are similar in structure to the ligaments that attach bones together across a joint.

Structure of a Muscle

A single muscle is fat in the middle and tapers towards the ends. The middle part, which gets fatter when the muscle contracts, is called the belly of the muscle. If you contract your biceps muscle in your upper arm you may feel it getting fatter in the middle. You may also notice that the biceps is attached at its top end to bones in your shoulder while at the bottom it is attached to bones in your lower arm. Notice that the bones at only one end move when you contract the biceps. This end of the muscle is called the insertion. The other end of the muscle, the origin, is attached to the bone that moves the least.

Antagonistic Muscles

Skeletal muscles usually work in pairs. When one contracts the other relaxes and vice versa. Pairs of muscles that work like this are called antagonistic muscles. For example the muscles in the upper forearm are the biceps and triceps. Together they bend the elbow. When the biceps contracts (and the triceps relaxes) the lower forearm is raised and the angle of the joint is reduced. This kind of movement is called flexion. When the triceps is contracted (and the biceps relaxes), the angle of the elbow increases. The term for this movement is extension.

When you or animals contract skeletal muscle it is a voluntary action. For example, you make a conscious decision to walk across the room, raise the spoon to your mouth or smile. There is however, another way in which contraction of muscles attached to the skeleton happens that is not under voluntary control. This is during a reflex action, such as jerking your hand away from the hot stove you have touched by accident.

Cardiovascular System/Blood

Blood is a unique fluid containing cells that is pumped by the heart around the body of animals in a system of pipes known as the circulatory system. It carries oxygen and nutrients to the cells of the body and removes waste products like carbon dioxide from them. Blood is also important for keeping conditions in the body constant, in other words for maintaining homeostasis. It helps keep the acidity or pH stable and helps maintain a constant temperature in the body. Blood also has an important role in defending the body against disease.

A simple way to find out what is in blood is to remove a small amount from an animal and place it in a tube with a substance that prevents it from clotting (an anticoagulant). If you leave the tube to stand for a few hours you will find that it settles out into two layers. The top layer consists of a light yellow fluid, the plasma, and the bottom layer consists of red blood cells (RBCs). If you look very carefully you can also see a thin beige-coloured layer in between these two layers. This consists of the white blood cells (WBCs).

The above procedure is usually done more rapidly by placing the blood sample in a centrifuge for a few minutes. This machine acts like a super spin drier rotating about 10,000 times a minute and packing the heavier particles (red blood cells) at the bottom of the tube. The sample that results is called the packed cell volume (P.C.V.) or haematocrit. It is a very useful measurement of the concentration of red blood cells in the blood. For most animals the packed cell volume is in the range 30-45%. If it is lower than this it means that the concentration of red blood cells is low and the animal is anaemic. If the reading is above this range it may mean the animal is dehydrated. Animals that live at high altitudes also have high P.C.V.s to compensate for the low oxygen concentration there.

Plasma

Plasma consists of water (91%) in which many substances are dissolved. These dissolved substances include:

- salts (or electrolytes)
- proteins
- nutrients
- waste products
- dissolved gases (mainly carbon dioxide)
- and other chemicals like hormones.

Salts in Plasma: Salts in the plasma are in the form of ions or electrolytes which include sodium, potassium, calcium, chloride, phosphate and bicarbonate. Plasma transports these ions to where they are needed e.g. calcium required by the bones, they also help keep the osmotic pressure and acid-base balance (pH) of the blood within the required levels.

Blood Proteins

The proteins in the blood plasma are large molecules with important functions. Some contribute to the osmotic pressure and the viscosity (thickness) of the blood, and so help keep the blood volume and pressure stable. Others act as antibodies that attack bacteria and viruses, and yet others are important in blood clotting. Nutrients that are absorbed from the gut and transported to the cells in the plasma include amino acids, glucose, fatty acids and vitamins. Waste products include urea from the breakdown of proteins.

Red Blood Cells

Red blood cells are also known as RBCs or erythrocytes. They are what make blood red. When you look at a blood smear through a microscope, as you will in one of the practical classes, you will see that RBCs are by far the most common cells in the blood. If you focus on an individual RBC, you will see that they are shaped like discs or doughnuts with a thin central portion surrounded by a fatter margin. This shape has all sorts of advantages, one being that enables the cells to fold up and pass along the narrowest blood capillaries.

Front and side	Red blood cell	Red blood cells as they
view of a red blood cell	cut in half	appear in a blood clot

Figure 33: Red blood cells or erythrocytes

The mature RBCs of mammals have neither nucleus nor other organelles and can be thought of as sacks of haemoglobin. Haemoglobin is a red coloured protein containing iron, which joins with oxygen so the blood can transport it to body cells. RBCs are made continuously in the bone marrow and live about 120 days. They are then destroyed in the liver and spleen and the molecules they are made from recycled to make new RBCs. Anaemia results if the rate at which RBCs are destroyed exceeds the rate at which RBC'c are produced.

Note that if you happen to look at bird's, reptiles, frogs or fishes blood down the microscope you will see that these vertebrates all have RBCs with a central nucleus.

White Blood Cells

White blood cells or leucocytes are far less numerous than red blood cells. In fact there is only about one white cell for every 1000 red blood cells. Rather than being white, they are actually colourless as they contain no haemoglobin although unlike RBCs they do have a nucleus. If you make a blood smear and look at it under the microscope it is difficult to see the white blood cells at all. To make them visible you need to stain them with special dyes or stains. There are a variety of stains that can be used, but most dye the nucleus a dark purple or pink colour. The stains may also show up the granules present in the cytoplasm of some white blood cells. White blood cells are divided into two major groups depending on the shape of the nucleus and whether or not there are granules in the cytoplasm.

1. Granulocytes or polymorphonuclear leucocytes ("polymorphs" or "polys") have granules in the cytoplasm and a purple lobed nucleus. The most common (neutrophils) can squeeze out of capillaries and are involved in engulfing and destroying foreign invaders like bacteria. Some (eosinophils) combat allergies and increase in numbers during parasitic worm infections. Others (basophils) produce heparin that prevents the blood from clotting.

2. Agranulocytes or monomorphonuclear leucocytes have a large unlobed nucleus and no granules in the cytoplasm. There are two types of agranulocytes. The most numerous are lymphocytes that are concerned with immune responses. The second type is the monocyte that is the largest blood cell and is involved in engulfing bacteria etc. by phagocytosis.

Platelets

As well as red and white blood cells, the blood also contains small irregular shaped fragments of cells known as platelets. They are involved in the clotting of the blood.

Transport of Oxygen

The purpose of the haemoglobin in red blood cells is to carry oxygen from the lungs to the tissues. In fact it allows the blood to carry about 25 times more oxygen than it would be able to without any haemoglobin. When oxygen concentrations are high, as in the blood capillaries in the lungs, haemoglobin combines with oxygen to form a compound called oxyhaemoglobin. This compound is bright red and makes the oxygenated blood that spurts from a damaged artery its characteristic bright red colour. When the blood reaches the tissues where the oxygen concentrations are low, the oxygen separates from the haemoglobin and diffuses into the tissues. The haemoglobin in most veins has given up its oxygen and the blood is called deoxygenated blood. It is a purple-red colour.

Carbon Monoxide Poisoning

Carbon monoxide is a colourless, odourless gas found in car exhaust fumes and tobacco smoke. It combines with haemoglobin just like oxygen but does not let go. This means the haemoglobin molecules are not available to carry oxygen to the tissues and the animal or human suffocates. Carbon monoxide poisoning is often fatal but can be treated by giving the patient pure oxygen that slowly replaces the carbon monoxide.

Transport of Carbon Dioxide

Carbon dioxide is a waste gas produced by cells. It diffuses into the blood capillaries where it is carried to the lungs in the blood. Most is carried in the plasma as bicarbonate ions but a small amount is dissolved directly in the plasma and some combines with haemoglobin.

Transport of Other Substances

The blood carries water to the cells and organs as well as soluble food substances (sugars, amino acids, fatty acids and vitamins) and hormones dissolved in the plasma. These are delivered to the cells via the tissue fluid that surrounds them. Blood also picks up the waste products like carbon dioxide and urea from the cells and is important in distributing the heat produced in the liver and muscles all over the body.

Blood Clotting

The mechanism that causes the blood to clot is easily seen when you or your animals are injured. However, minor injuries occur all the time in areas that experience wear and tear like the intestine, the lungs and the skin. Without the clotting mechanism, animals would quickly bleed to death from minor injury and internal haemorrhage. This is what happens in animals and people with clotting disorders like haemophilia, as well as animals that are poisoned with rat poisons like warfarin.

Platelets are important in blood clotting. When blood vessels are damaged, substances released cause the blood platelets to disintegrate. This stimulates a complex chain of reactions, which causes the protein fibrinogen to be converted to fibrin. Fibrin forms a dense fibrous network over the wound preventing the escape of further blood. Calcium and vitamin K are essential for the clotting process and any deficiency of these may also lead to clotting problems.

Serum and Plasma

When blood clots it separates into the clot that contains most of the cells and platelets leaving behind a straw coloured fluid. This fluid is called serum. It looks just like plasma and is similar in composition except for one big difference. It doesn't contain fibrinogen, the protein that forms the clot.

Anticoagulants

Anticoagulants are substances that interfere with the clotting process. When blood is collected for transfusion or testing it is often important to prevent it clotting and there are a number of different anticoagulants you can use for this. Tubes containing the different anticoagulants are coded with different colours for easy recognition.

1. Heparin (colour code - green) is a natural anticoagulant produced by the white blood cells but it is also used routinely in the laboratory with samples to be tested for heavy metals like lead.
2. EDTA (colour code – lavender) is used for routine blood counts.
3. Fluoroxylate (colour code – grey) is used for biochemical tests for glucose.
4. Citrate (colour code – light blue) is used for the storage of large quantities of blood, such as used in transfusions.

Haemolysis

Haemolysis is the breakdown of the plasma membrane of red blood cells to release the haemoglobin. We have already met this process when discussing osmosis, for haemolysis often occurs when red blood cells are placed in a hypotonic solution and water flows in through the semi permeable plasma membrane to swell and eventually burst the cell. It is therefore important when collecting blood from an animal to make sure there is no water in the syringe or tube. Too much movement due to shaking the tube or sucking up the blood too vigorously can also break down the plasma membrane and cause haemolysis.

Blood Groups

If you have given blood recently you may know your blood group. It may be blood group O, A or B or even AB, the rarest group. Blood groups are the result of different molecules called antigens on the outside of red blood cells. These cause antibodies to be formed that attack viruses and bacteria. Knowledge of a person's blood group is important when giving transfusions because if blood of another incompatible blood group is given to a patient the red blood cells stick together and block the blood vessels and may lead to death.

Blood groups also exist in many animals. There are three blood groups in cats and great care has to be taken that the groups are compatible when transfusing exotic breeds. The situation is slightly different in dogs. They have a number of blood groups but there is usually no problem with the first blood transfusion a dog receives. However, this first transfusion sensitises the immune system so that a problem may arise with the second and subsequent transfusions.

Haemolysis can occur in the living animal when it is exposed to various poisons and toxins. This may happen when, for example, it eats a poisonous plant, is bitten by a snake or infected with bacteria that destroy red blood cells (haemolytic bacteria).

Blood Volume

Blood accounts for between 6-10% of the body weight of animals, varying with the species and the stage of life. Animals can not tolerate losses of greater than 3% of the total volume when the condition known as shock occurs.

Summary Blood

- The main functions of blood are transport of oxygen, food, waste products etc., the maintenance of homeostasis and defending the body from disease.
- Blood consists of fluid, plasma, in which red and white blood cells are suspended. The blood cells typically make up 30-45% of the blood volume.
- Plasma consists of water containing dissolved substances like proteins, nutrients and carbon dioxide.
- Red Blood Cells contain haemoglobin to transport oxygen.
- White Blood Cells defend the body from invasion. There are 2 kinds:
- Granular white cells include neutrophils, basophils and eosinophils. Neutrophils which destroy bacteria are the most numerous. Eosinophils are involved with allergies and parasitic infections.

- Non-granular white cells include lymphocytes that produce antibodies to attach bacteria and viruses and monocytes that engulf and destroy bacteria and viruses.
- Platelets are involved in blood clotting.

Cardiovascular System/The Heart

The heart is the pump that pushes the blood around the body in the blood vessels of the circulatory system. In fishes the blood only passes through the heart once on its way to the gills and then round the rest of the body. However, in mammals and birds that have lungs, the blood passes through the heart twice: once on its way to the lungs where it picks up oxygen and then through the heart again to be pumped all over the body. The heart is therefore two separate pumps, side by side.

The heart is situated in the thorax between the lungs and is protected by the rib cage. In some animals it is displaced slightly to the left-hand side. A tough membrane called the pericardium covers it. There is a narrow space between the pericardium and the heart that is filled with a liquid that acts as a lubricant.

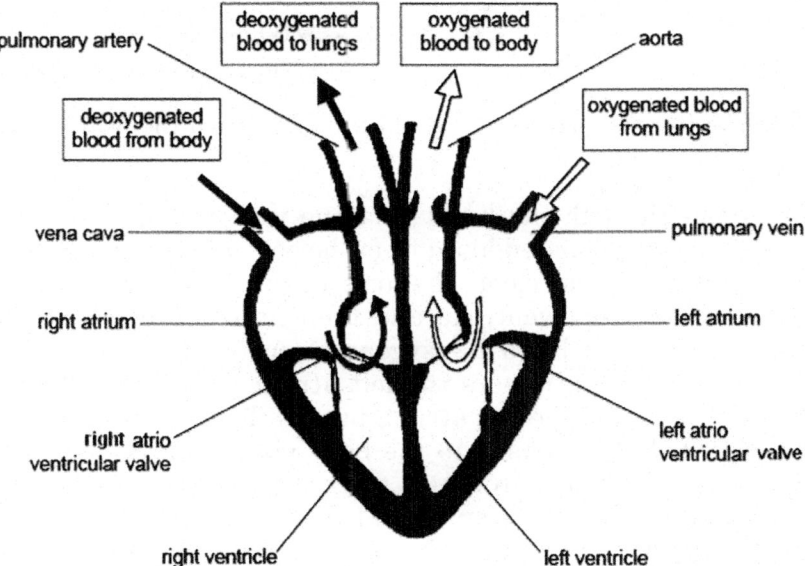

Figure 34: Simplified diagram of the internal structure of the mammalian heart

The heart of mammals is a hollow bag made of cardiac muscle. The cavity inside the heart is divided into 4 chambers. The chambers on the right side are completely separate from the chambers on the left side. The two upper chambers are thin walled, and are called the atria (or auricles). The two lower chambers are thick walled and are called the ventricles.

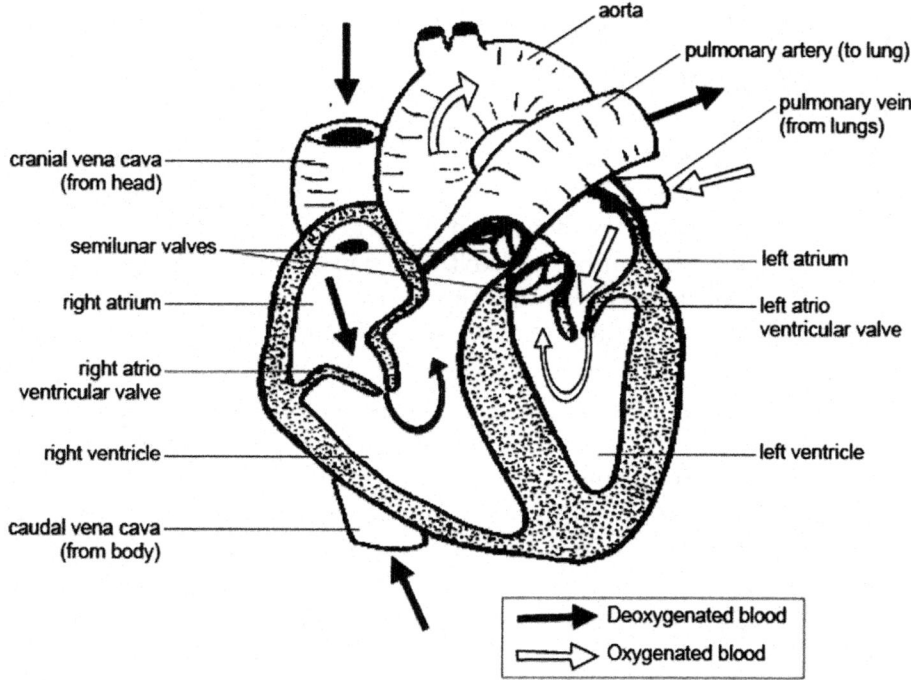

Figure 35: The internal structure and blood flow through the heart

The Heart

Blood flows through the heart in a one way system. The right atrium receives deoxygenated blood from the body via the largest vein in the body called the vena cava. The contraction of the atrium pumps the blood into the right ventricle and then into the lungs via the pulmonary artery. The blood is oxygenated in the lungs and then returns to the heart and enters the left atrium via the pulmonary vein. The contraction of the left atrium pumps the blood into the left ventricle, which then pumps it to the body via the aorta. The wall of the left ventricle is usually much thicker than that of the right ventricle because it has to pump the blood to the end of the digits and tip of the tail while the right ventricle only has to pump the blood to the nearby lungs.

Valves

Valves are flaps of tissue that stop blood flowing backwards and so control the direction of blood flow in the heart. There are two kinds of valves in the heart. The first kind is the massive valves between the atria and the ventricles, the atrio-ventricular valves, (AV valves) that prevent blood in the ventricles from flowing back into the atria. The flaps of these valves are attached to the walls of the ventricles by tendons. These make them look somewhat like parachutes.

The second kind of valve is pocket shaped flaps of tissue called the semilunar (half moon) valves. They are called the pulmonary and aortic valves and found at the back of the pulmonary artery and aorta respectively.

The Heartbeat

The heartbeat consists of alternating contractions and relaxations of the heart. It you listen to the heart with a stethoscope you hear the sounds often described as "lubb-dupp".

There are four stages to each heartbeat:

1. Each atrium relaxes so that blood can enter. Blood flows from the body via the vena cava into the :right atrium. At the same time, blood flows from the lungs via the pulmonary vein into the left :atrium.

2. The atrio-ventricular valves open and both ventricles relax. The atria contract and blood flows from the right atrium into the right ventricle and from the left atrium into the left ventricle.

3. The ventricles contract and the atrio-ventricular valves snap shut to stop blood flowing back into the atria. This is the first sound ("lubb") of the heartbeat that can be heard with a stethoscope.

4. The semi-lunar valves open and blood is pumped out of the right ventricle to the lungs. At the same time, blood is pumped out of the left ventricle into the aorta and so to the rest of the body. When the ventricles stop contracting the semi-lunar valves snap shut to stop blood flowing backwards.

This is the second sound ("dupp") of the heartbeat. Blood flows into the atria again as they relax and the cycle is repeated.

When a valve is damaged and fails to close completely some blood may flow backwards after each heartbeat. A trained veterinarian hears this with a stethoscope as a "heart murmur".

The period of the heart beat when the ventricles are contracting and sending a wave of blood down the pulmonary artery and aorta is called systole. The period when the ventricles are relaxing is called diastole.

Cardiac Muscle

The walls of the heart consist of cardiac muscle, a special kind of muscle only found in the heart. The cells of cardiac muscle form a branching network of separating and rejoining fibres which allows nerve impulses to travel through the tissue. Heart muscle needs lots of energy to function so it is well supplied with mitochondria and requires a good supply of oxygen. This is provided by the coronary arteries.

Control of the Heartbeat

The cardiac muscle of the walls of the heart contracts of its own accord. This can be demonstrated by the rather macabre experiment in which a small portion of heart muscle is removed and placed in a solution that is similar to blood. The tissue will continue to contract and relax for a time. In the normal functioning heart the pacemaker acts rather like the conductor of an orchestra and superimposes a unified beat upon the heart as a whole. The pacemaker is situated in the wall of the right atrium. The rate at which the heart beats is modified by a part of the brain called the medulla oblongata, and by the hormone adrenalin, which speeds up the heartbeat.

The Coronary Vessels

Although oxygenated blood passes through some of the chambers of the heart it can not supply the muscle of the heart walls with the oxygen and nutrients it needs. Special arteries called the coronary arteries do this.

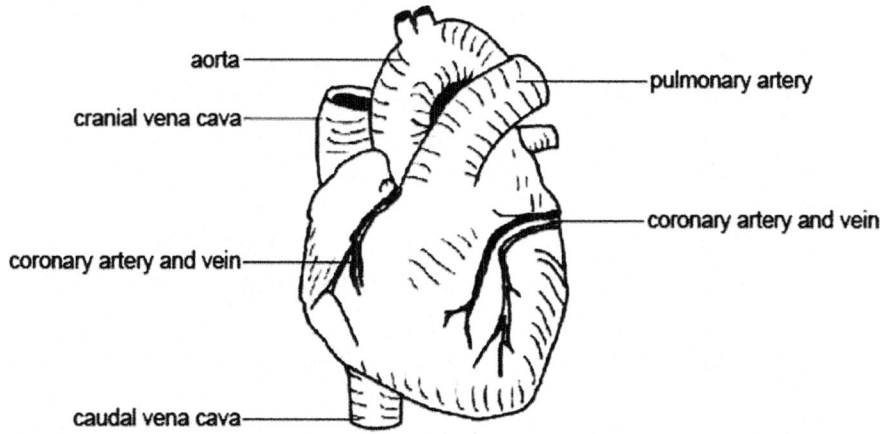

Figure 36: The heart showing coronary vessels

These two arteries arise from the aorta and branch through the heart to deliver oxygen and nutrients to the cardiac muscles and collect carbon dioxide and wastes. Coronary veins return the blood to right hand side of the heart. Some of these vessels can be seen on the outside surface of the heart. Sometimes fatty deposits on the inside wall of the coronary artery block the blood flow to the heart muscle. If the obstruction is severe enough to damage the heart muscle due to inadequate blood supply a "heart attack" can result.

Summary

- The heart is situated in the thorax between the lungs
- The heart is a hollow bag made of cardiac muscle. It is divided into four chambers (right and left atria and right and left ventricles).
- Valves stop blood flowing backwards. The right and left atrio-ventricular valves prevent blood in the ventricles from flowing back into the atria. The semilunar valves at the entrance of the pulmonary artery and aorta prevent blood flowing back into the ventricles. The closing of the valves produces the heart sounds heard with a stethoscope.
- There are 4 stages to the heart beat. 1. blood flows into the right and left atria. 2. The atria contract and blood flows into the ventricles. 3. The ventricles contract and the closing of the atrio-ventricular valves produces the first heart sound. 4. Blood flows to the lungs and body and when the ventricles stop contracting the closing of the semilunar valves produces the second heart sound.
- The coronary arteries supply the heart muscle with oxygenated blood.

Chapter 2

Anatomy and Physiology of Animals Sensitive Organs

Cardiovascular System/Blood Circulation

The circulatory system is the continuous system of tubes through which the blood is pumped around the body. It supplies the tissues with their requirements and removes waste products. In mammals and birds the blood circulates through two separate systems - the first from the heart to the lungs and back to the heart again (the pulmonary circulation) and the second from the heart to the head and body and back again. The tubes through which the blood flows are the arteries, capillaries and veins. The heart pumps blood into arteries that carry it away from the heart. The arteries divide into very thin vessels called capillaries that form a network between the cells of the body. The capillaries then join up again to make veins that return the blood to the heart.

Arteries

Arteries carry blood away from the heart. They have thick elastic walls that stretch and can withstand the surges of high pressure blood caused by the heartbeat (the pulse). The arteries divide into smaller vessels called arterioles. The hole down the centre of the artery is called the lumen. There are three layers of tissue in the walls of an artery. It is lined with squamous epithelial cells. The middle layer is the thickest layer. It made of elastic fibres and smooth muscle to make it stretchy. The outer fibrous layer protects the artery. The pulse is only felt in arteries.

The Pulse

The pulse is the spurt of high pressure blood that passes along the aorta and arteries when the left ventricle contracts. As the pulse of blood passes along an artery the elastic walls stretch. When the pulse has passed the walls contract and this helps push the blood along. The pulse is easily felt at certain places where an artery passes near the surface of the body. It is strongest near the heart and becomes weaker as it travels away from the heart. The pulse disappears altogether in the capillaries.

Capillaries

Arterioles divide repeatedly to form a network penetrating between the cells of all tissues of the body. These small vessels are called capillaries. The walls are only one cell thick and some capillaries are so narrow that red blood cells have to fold up to pass through them. Capillaries form networks in tissues called capillary beds. The capillary networks in capillary beds are so dense that no living cell is far from its supply of oxygen and food.

Note: All arteries carry oxygenated blood except for the pulmonary artery that carries deoxygenated blood to the lungs.

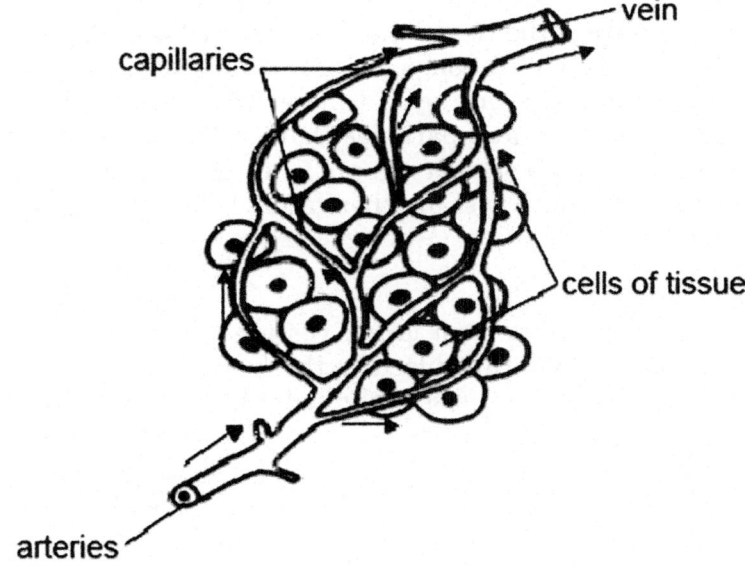

Figure 1: A capillary bed

The Formation of Tissue Fluid and Lymph

The thin walls of capillaries allow water, some white blood cells and many dissolved substances to diffuse through them. These form a clear fluid called tissue fluid (or extracellular fluid or interstitial fluid) that surrounds the cells of the tissues. The tissue fluid allows oxygen and nutrients to pass from the blood to the cells and carbon dioxide and other waste products to be removed from the tissues.

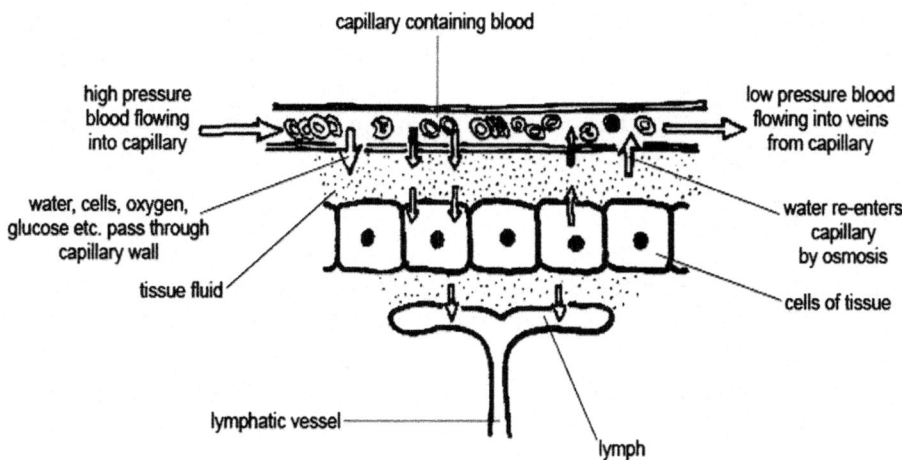

Figure 2: The formation of tissue fluid and lymph from blood

Some tissue fluid finds its way back into the capillaries and some of it flows into the blind-ended lymphatic vessels that form a network in the tissues. Once the tissue fluid has entered the lymphatics it is called lymph although its composition remains the same. The lymph vessels have walls that are even thinner than the capillaries. This means that molecules and particles that are larger than those that can pass into the blood stream e.g. cancer cells and bacteria can enter the lymphatic system. These are then filtered out as the lymph passes through lymph nodes.

Veins

Capillaries unite to form larger vessels called venules that join to form veins. Veins return blood to the heart and since blood that flows in veins has already passed through the fine capillaries, it flows slowly with no pulse and at low pressure. For this reason veins have thinner walls than arteries although there have the same three layers in them as arteries. As there is no pulse in veins, the blood is squeezed along them by the contraction of the skeletal muscles that lay alongside them. Veins also have valves in them that prevent blood flowing backwards.

The pulmonary vein that carries oxygenated blood from the lungs to the left atrium of the heart is an exception.

Regulation of Blood Flow

The flow of blood along arteries, arterioles and capillaries is not constant but can be controlled depending upon the requirements of the body. For example more blood is directed to the skeletal muscles, brain or digestive system when they are active. Regulation of the blood flow to the arterioles of the skin is also important in controlling body temperature. The size of the vessels is adjusted by the contraction or relaxation of smooth muscle fibres in their walls.

Oedema and Fluid Loss

Oedema is the swelling of the tissues due to the accumulation of tissue fluid. This may occur because the tissue fluid is prevented from returning to the bloodstream and accumulates in the tissues. This may be caused by physical inactivity (e.g. long car or plane trips in humans) or because of imbalances in the proteins in the blood. This is what causes the "potbelly" of the malnourished child or worm-infested puppy.

Loss of body fluid can be caused not only by drinking insufficient liquid but also through diarrhea and vomiting or sudden loss of blood due to haemorrhage. The effect is to reduce the volume of the blood which decreases the blood pressure. This could be dangerous because the supply of adequate blood to the brain depends upon maintaining the blood pressure at a constant level.

To compensate for the loss of fluid various mechanisms come into play. First of all the blood vessels contract in order to try and maintain the pressure. Then, since the loss of fluid tends to make the blood more concentrated and increases its osmotic pressure, fluid is drawn into the blood from the tissues by osmosis.

The Spleen

The spleen is situated near the stomach. It has a rich blood supply and acts as a reservoir of red blood cells. When there is a sudden loss of blood, as happens when a haemorrhage occurs, the spleen contracts to release large numbers of red blood cells into the circulation. The spleen also destroys old red cells and makes new lymphocytes but it is not an essential organ because its removal in adult life seems to cause few problems. In the foetus, the spleen makes both red and white cells.

Important Blood Vessels of The Systemic (Body) Circulation

Blood is pumped out into the body via the main artery, the aorta. This takes the blood to the head, the limbs and all the body organs. After passing through a network of fine capillaries, the blood is returned to the heart in the largest vein, the vena cava.

Arteries and veins to and from many organs often run alongside each other and have the same name e.g. the renal artery and vein serve the kidney, the femoral artery and vein serve the hind limbs and the subclavian artery and vein serve the forelimbs. However, blood to the head passes along the carotid artery and returns to the cranial vena cava via the jugular vein.

One variation on this arrangement is found in the blood vessels that serve the digestive tract. A variety of arteries take blood from the aorta to the intestines but blood from the intestines is carried by the hepatic portal vein to the liver where the digested food can be processed. This vessel is unlike others in that it transports blood from one organ to another rather than to or from the heart like arteries or veins.

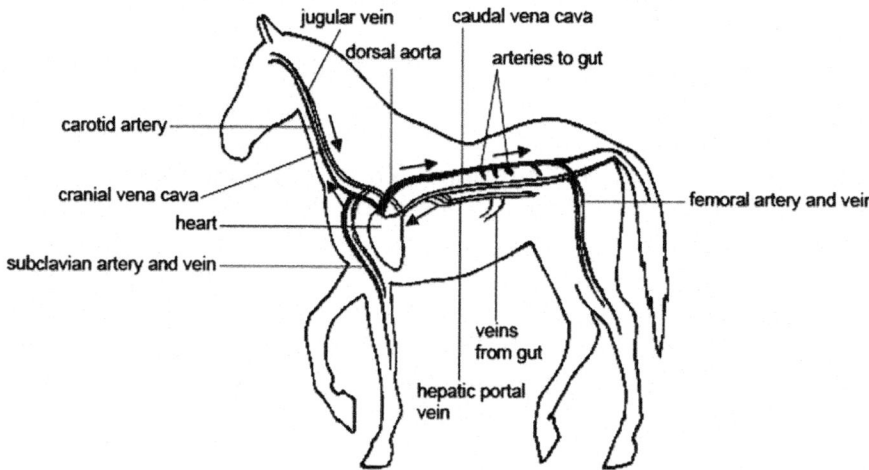

Figure 3: The main arteries and veins of the horse

Blood Pressure

The blood pressure is the pressure of the blood against the walls of the main arteries. The pressure is highest as the pulse produced by the contraction of the left ventricle passes along the artery. This is known as the systolic pressure. Pressure is much lower between pulses. This is known as the diastolic pressure. Blood pressure is measured in millimetres of mercury. A blood pressure that is higher than expected is known as hypertension while a pressure lower than expected is known as hypotension.

- The circulatory system is double with the blood passing through the heart twice.

- Arteries carry blood away from the heart. They have thick elastic walls that stretch and can withstand the high pressure of the pulse.

- Capillaries are small, thin walled vessels that form a network between the cells of the tissues.

- Veins return low pressure blood to the heart. They have thinner walls than arteries.

- The pulse is the spurt of high pressure blood that passes along the arteries when the left ventricle contracts. It can be felt where arteries pass close to the body surface.

- Tissue fluid is the clear fluid that leaks from the capillaries and surrounds the cells of the tissues. Lymph forms when tissue fluid enters lymphatics.

- Important blood vessels include the vena cava, aorta, pulmonary artery, carotid artery, jugular vein, renal artery and vein and hepatic portal vessel.

Respiratory System

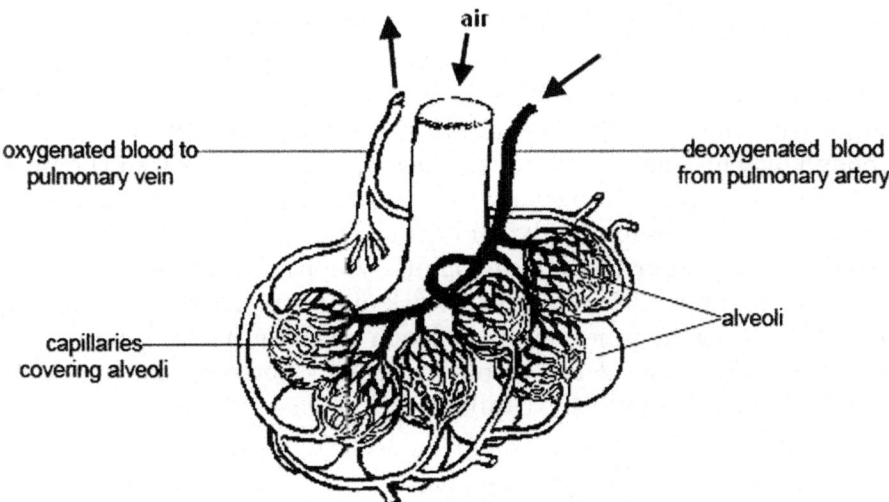

air

oxygenated blood to pulmonary vein

deoxygenated blood from pulmonary artery

alveoli

capillaries covering alveoli

Figure 4: Alveoli with blood supply

Animals require a supply of energy to survive. This energy is needed to build large molecules like proteins and glycogen, make the structures in cells, move chemicals through membranes and around cells, contract muscles, transmit nerve impulses and keep the body warm. Animals get their energy from the large molecules that they eat as food. Glucose is often the energy source but it may also come from other carbohydrates, as well as fats and protein. The energy is made by the biochemical process known as cellular respiration that takes place in the mitochondria inside every living cell.

The overall reaction can be summarised by the word equation given below.

Charbohydrate Food (glucose) + Oxygen = Carbon Dioxide + Water + energy

As you can see from this equation, the cells need to be supplied with oxygen and glucose and the waste product, carbon dioxide, which is poisonous to cells, needs to be removed. The way the digestive system provides the glucose for cellular respiration will be described in Chapter 5, but here we are only concerned with the two gases, oxygen and carbon dioxide, that are involved in cellular respiration. These gases are carried in the blood to and from the tissues where they are required or produced. Oxygen enters the body from the air (or water in fish)and carbon dioxide is usually eliminated from the same part of the body. This process is called gas exchange. In fish gas exchange occurs in the gills, in land dwelling vertebrates lungs are the gas exchange organs and frogs use gills when they are tadpoles and lungs, the mouth and the skin when adults.

Mammals (and birds) are active and have relatively high body temperatures so they require large amounts of oxygen to provide sufficient energy through cellular respiration. In order to take in enough oxygen and release all the carbon dioxide produced they need a very large surface area over which gas exchange can take place. The many minute air sacs or alveoli of the lungs provide this. When you look at these under the microscope they appear rather like bunches of grapes covered with a dense network of fine capillaries. A thin layer of water covers the inner surface of each alveolus. There is only a very small distance -just 2 layers of thin cells - between the air in the alveoli and the blood in the capillaries. The gases pass across this gap by diffusion.

Diffusion And Transport of Oxygen

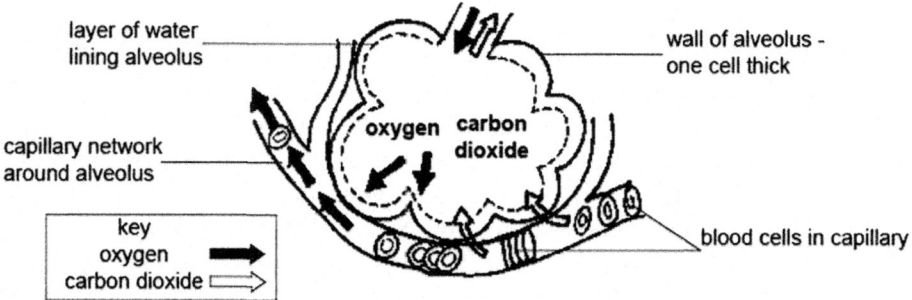

Figure 5: Cross section of an alveolus

The air in the alveoli is rich in oxygen while the blood in the capillaries around the alveoli is deoxygenated. This is because the haemoglobin in the red blood cells has released all the oxygen it has been carrying to the cells of the body. Oxygen diffuses from high concentration to low concentration. It therefore crosses the narrow barrier between the alveoli and the capillaries to enter the blood and combine with the haemoglobin in the red blood cells to form oxyhaemoglobin.

The narrow diametre of the capillaries around the alveoli means that the blood flow is slowed down and that the red cells are squeezed against the capillary walls. Both of these factors help the oxygen diffuse into the blood.

When the blood reaches the capillaries of the tissues the oxygen splits from the haemoglobin molecule. It then diffuses into the tissue fluid and then into the cells.

Diffusion and Transport of Carbon Dioxide

Blood entering the lung capillaries is full of carbon dioxide that it has collected from the tissues. Most of the carbon dioxide is dissolved in the plasma either in the form of sodium bicarbonate or carbonic acid. A little is transported by the red blood cells. As the blood enters the lungs the carbon dioxide gas diffuses through the capillary and alveoli walls into the water film and then into the alveoli. Finally it is removed from the lungs during breathing out.

The Air Passages

When air is breathed in it passes from the nose to the alveoli of the lungs down a series of tubes. After entering the nose the air passes through the nasal cavity, which is lined with a moist membrane that adds warmth and moisture to the air as it passes. The air then flows through the pharynx or throat, a passage that carries both food and air, to the larynx where the voice-box is located. Here the passages for food and air separate again. Food must pass into the oesophagus and the air into the windpipe or trachea. To prevent food entering this, a small flap of tissue called the epiglottis closes the opening during swallowing. A reflex that inhibits breathing during swallowing also (usually) prevents choking on food.

The trachea is the tube that ducts the air down the throat. Incomplete rings of cartilage in its walls help keep it open even when the neck is bent and head turned. The fact that acrobats and people that tie themselves in knots doing yoga still keep breathing during the most contorted manoeuvres shows how effective this arrangement is. The air passage now divides into the two bronchi that take the air to the right and left lungs before dividing into smaller and smaller bronchioles that spread throughout the lungs to carry air to the alveoli. Smooth muscles in the walls of the bronchi and bronchioles adjust the diametre of the air passages.

The tissue lining the respiratory passages produces mucus and is covered with minute hairs or cilia. Any dust that is breathed into the respiratory system immediately gets entangled in the mucous and the cilia move it towards the mouth or nose where it can be coughed up or blown out.

The Lungs and the Pleural Cavities

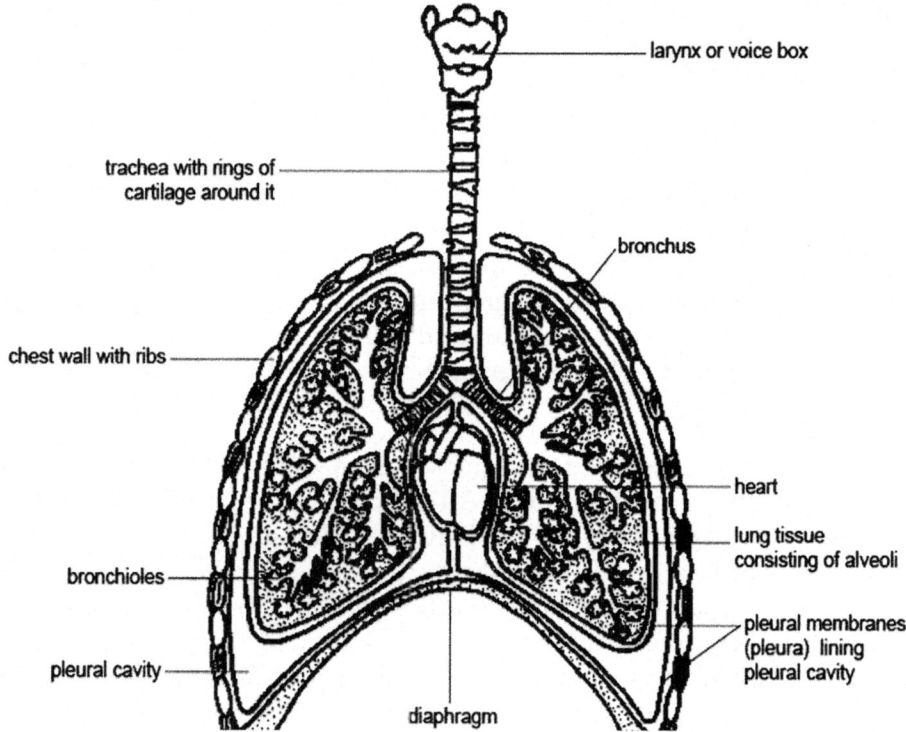

Figure 6: The respiratory system

The lungs fill most of the chest or thoracic cavity, which is completely separated from the abdominal cavity by the diaphragm. The lungs and the spaces in which they lie (called the pleural cavities) are covered with membranes called the pleura. There is a thin film of fluid between the two membranes. This lubricates them as they move over each other during breathing movements.

Collapsed Lungs

The pleural cavities are completely airtight with no connection with the outside and if they are punctured by accident (a broken rib will often do this), air rushes in and the lung collapses. Separating the two lungs is a region of tissue that contains the oesophagus, trachea, aorta, vena cava and lymph nodes. This is called the mediastinum. In humans and sheep it separates the cavity completely so that puncturing one pleural cavity leads to the collapse of only one lung. In dogs, however, this separation is incomplete so a puncture results in a complete collapse of both lungs.

Breathing

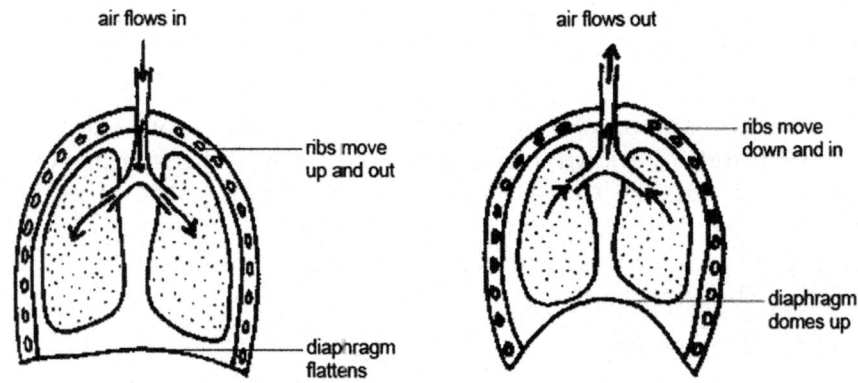

Figure 7: (a) Inspiration; (b) Expiration

The process of breathing moves air in and out of the lungs. Sometimes this process is called respiration but it is important not to confuse it with the chemical process, cellular respiration, that takes place in the mitochondria of cells. Breathing is brought about by the movement of the diaphragm and the ribs.

Inspiration

The diaphragm is a thin sheet of muscle that completely separates the abdominal and thoracic cavities. When at rest it domes up into the thoracic cavity but during breathing in or inspiration it flattens. At the same time special muscles in the chest wall move the ribs forwards and outwards. These movements of both the diaphragm and the ribs cause the volume of the thorax to increase. Because the pleural cavities are airtight, the lungs expand to fill this increased space and air is drawn down the trachea into the lungs.

Expiration

Expiration or breathing out consists of the opposite movements. The ribs move down and in and the diaphragm resumes its domed shape so the air is expelled. Expiration is usually passive and no energy is required (unless you are blowing up a balloon).

Lung Volumes

As you sit here reading this just pay attention to your breathing. Notice that your in and out breaths are really quite small and gentle (unless you have just rushed here from somewhere else!). Only a small amount of the total volume that your lungs hold is breathed in and out with each breath.

This kind of gentle "at rest" breathing is called tidal breathing and the volume breathed in or out (they should be the same) is the tidal volume. Sometimes people want to measure the volume of air inspired or expired during a minute of this normal breathing. This is called the minute volume.

It could be estimated by measuring the volume of one tidal breath and then multiplying that by the number of breaths in a minute. Of course it is possible to take a deep breath and breathe in as far as you can and then expire as far as possible. The volume of the air expired when a maximum expiration follows a maximum inspiration is called the vital capacity.

Composition of Air

The air animals breathe in consists of 21% oxygen and 0.04% carbon dioxide. Expelled air consists of 16% oxygen and 4.4% carbon dioxide. This means that the lungs remove only a quarter of the oxygen contained in the air. This is why it is possible to give someone (or an animal) artificial respiration by blowing expired air into their mouth.

Breathing is usually an unconscious activity that takes place whether you are awake or asleep, although, humans at least, can also control it consciously. Two regions in the hindbrain called the medulla oblongata and pons control the rate of breathing. These are called respiratory centres. They respond to the concentration of carbon dioxide in the blood. When this concentration rises during a bout of activity, for example, nerve impulses are automatically sent to the diaphragm and rib muscles that increase the rate and the depth of breathing. Increasing the rate of breathing also increases the amount of oxygen in the blood to meet the needs of this increased activity.

The Acidity of the Blood and Breathing

The degree of acidity of the blood (the acid-base balance) is critical for normal functioning of cells and the body as a whole. For example, blood that is too acidic or alkaline can seriously affect nerve function causing a coma, muscle spasms, convulsions and even death. Carbon dioxide carried in the blood makes the blood acidic and the higher the concentration of carbon dioxide the more acidic it is. This is obviously dangerous so there are various mechanisms in the body that bring the acid-base balance back within the normal range. Breathing is one of these homeostatic mechanisms. By increasing the rate of breathing the animal increases the amount of dissolved carbon dioxide that is expelled from the blood. This reduces the acidity of the blood.

Breathing In Birds

Birds have a unique respiratory system that enables them to respire at the very high rates necessary for flight. The lungs are relatively solid structures that do not change shape and size in the same way as mammalian lungs do. Tubes run through them and connect with a series of air sacs situated in the thoracic and abdominal body cavities and some of the bones. Movements of the ribs and breastbone or sternum expand and compress these air sacs so they act rather like bellows and pump air through the lungs. The evolution of this extremely efficient system of breathing has enabled birds to migrate vast distances and fly at altitudes higher than the summit of Everest.

- Animals need to breathe to supply the cells with oxygen and remove the waste product carbon dioxide.
- The lungs are situated in the pleural cavities of the thorax.
- Gas exchange occurs in the alveoli of the lungs that provide a large surface area. Here oxygen diffuses from the alveoli into the red blood cells in the capillaries that surround the alveoli. Carbon dioxide, at high concentration in the blood, diffuses into the alveoli to be breathed out.
- Inspiration occurs when muscle contraction causes the ribs to move up and out and the diaphragm to flatten. These movements increase the volume of the pleural cavity and draw air down the respiratory system into the lungs.
- The air enters the nasal cavity and passes to the pharynx and larynx where the epiglottis closes the opening to the lungs during swallowing. The air passes down the trachea kept open

by rings of cartilage to the bronchi and bronchioles and then to the alveoli.

- Expiration is a passive process requiring no energy as it relies on the relaxation of the muscles and recoil of the elastic tissue of the lungs.

- The rate of breathing is determined by the concentration of carbon dioxide in the blood. As carbon dioxide makes blood acidic, the rate of breathing helps control the acid/base balance of the blood.

- The cells lining the respiratory passages produce mucus which traps dust particles, which are wafted into the nose by cilia.

Lymphatic System

When tissue fluid enters the small blind-ended lymphatic capillaries that form a network between the cells it becomes lymph. Lymph is a clear watery fluid that is very similar to blood plasma except that it contains large numbers of white blood cells, mostly lymphocytes. It also contains protein, cellular debris, foreign particles and bacteria. Lymph that comes from the intestines also contains many fat globules following the absorption of fat from the digested food into the lymphatics (lacteals) of the villi. From the lymph capillaries the lymph flows into larger tubes called lymphatic vessels. These carry the lymph back to join the blood circulation.

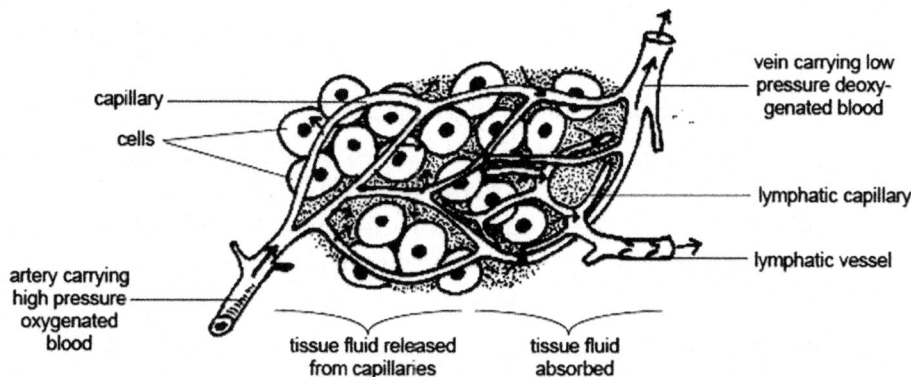

Figure 8: A capillary bed with lymphatic capillaries

Lymphatic Vessels

Lymphatic vessels have several similarities to veins. Both are thin walled and return fluid to the right hand side of the heart. The movement of the fluid in both is brought about by the contraction of the muscles that surround them and both have valves to prevent backflow. One important difference is that lymph passes through at least one lymph node or gland before it reaches the blood system. These filter out used cell parts, cancer cells and bacteria and help defend the body from infection. Lymph nodes are of various sizes and shapes and found throughout the body and the more important ones are shown. They consist of lymph tissue surrounded by a fibrous sheath. Lymph flows into them through a number of incoming vessels. It then trickles through small channels where white cells called macrophages (derived from monocytes) remove the bacteria and debris by engulfing and digesting them. The lymph then leaves the lymph nodes through outgoing vessels to continue its journey towards the heart where it rejoins the blood circulation.

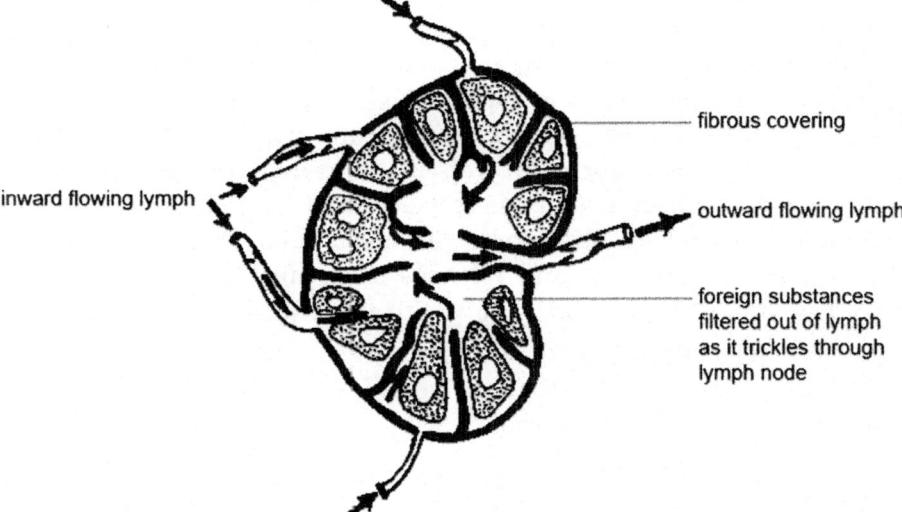

Figure 9: A lymph node

As well as filtering the lymph, lymph nodes produce the white cells known as lymphocytes. Lymphocytes are also produced by the thymus, spleen and bone marrow. There are two kinds of lymphocyte. The first attach invading micro organisms directly while others produce antibodies that circulate in the blood and attack them.

The function of the lymphatic system can therefore be summarised as transport and defence. It is important for returning the fluid and proteins that have escaped from the blood capillaries to the blood system and is also responsible for picking up the products of fat digestion in the small intestine. Its other essential function is as part of the immune system, defending the body against infection.

Problems with Lymph Nodes and the Lymphatic System

During infection of the body the lymph nodes often become swollen and tender because of their increased activity. This is what causes the swollen 'glands' in your neck during throat infections, mumps and tonsillitis. Sometimes the bacteria multiply in the lymph node and cause inflammation. Cancer cells may also be carried to the lymph nodes and then transported to other parts of the body where they may multiply to form a secondary growth or metastasis. The lymphatic system may therefore contribute to the spread of cancer. Inactivity of the muscles surrounding the lymphatic vessels or blockage of these vessels causes tissue fluid to 'back up' in the tissues resulting in swelling or oedema.

Other Organs of the Lymphatic System

The spleen is an important part of the lymphatic system. It is a deep red organ situated in the abdomen caudal to the stomach. It is composed of two different types of tissue. The first type makes and stores lymphocytes, the cells of the immune system. The second type of tissue destroys worn out red blood cells, breaking down the haemoglobin into iron, which is recycled, and waste products that are excreted. The spleen also stores red blood cells. When severe blood loss occurs, it contracts and releases them into the circulation.

The thymus is a large pink organ lying just under the sternum (breastbone) just cranial to the heart. It has an important function processing lymphocytes so they are capable of recognising and attacking foreign invaders like bacteria.

Other lymph organs are the bone marrow of the long bones where lymphocytes are produced and lymph nodules, which are like tiny lymph nodes. Large clusters of these are found in the wall of the small intestine (called Peyer's Patches) and in the tonsils.

The Gut and Digestion

Plant cells are made of organic molecules using energy from the sun. This process is called photosynthesis. Animals rely on these ready-made organic molecules to supply them with their food. Some animals (herbivores) eat plants; some (carnivores) eat the herbivores.

Herbivores

Although herbivores eat plant material they have a basic problem. Their own digestive enzymes do not break down the large cellulose molecules in the plant cell walls. Micro-organisms like bacteria, on the other hand, can break them down. Therefore herbivores employ micro-organisms to do the job for them.

There are two types of herbivore:

- The first, ruminants like cattle, sheep and goats, house these bacteria in a special compartment in the enlarged stomach called the rumen.

- The second group has an enlarged large intestine and caecum, called a functional caecum, occupied by cellulose digesting micro-organisms. These non-ruminant herbivores include the horse, rabbit and rat.

Even with the help of microorganisms, eating plants still poses problems. Plants are not a good source of nutrients, and herbivores have to eat large quantities of food to obtain all they require. Herbivores like cows, horses and rabbits typically spend much of their day feeding. To give the micro- organisms access to the cellulose molecules, the plant cell walls need to be broken down. This is why herbivores have teeth that are adapted to crush and grind. Their guts also tend to be lengthy and the food takes a long time to pass through it.

Eating plants does have some advantages. Plants are immobile so herbivores normally have to spend little energy collecting them. This contrasts with the other main group of animals- the carnivores that often have to chase their prey.

Carnivores

Carnivorous animals like those in the cat and dog families, bears, seals, crocodiles and birds of prey catch and eat other animals. They often have to use large amounts of energy finding, stalking, catching and killing their prey.

However, they are rewarded by the fact that meat provides a very concentrated source of nutrients. Carnivores in the wild therefore tend to eat distinct meals often with long and irregular intervals between them. Time after feeding is spent digesting and absorbing the food.

The guts of carnivores are usually shorter and less complex than those of herbivores because meat is easier to digest than plant material. Carnivores usually have teeth that are specialised for dealing with flesh, gristle and bone. They have sleek bodies, strong, sharp claws and keen senses of smell, hearing and sight. They are also often cunning, alert and have an aggressive nature.

Omnivores

Many animals feed on both animal and vegetable material – they are omnivorous. Most primates including humans belong to this category as do pigs and rats. Their food is diverse, ranging from plant material to animals they have either killed themselves or scavenged from other carnivores. Omnivores lack the specialised teeth and guts of carnivores and herbivores but are often highly intelligent and adaptable reflecting their varied diet.

Treatment of Food

Whether an animal eats plants or flesh, the carbohydrates, fats and proteins in the food it eats are generally giant molecules. These need to be split up into smaller ones before they can pass into the blood and enter the cells to be used for energy or to make new cell constituents.

For example:

*   Carbohydrates like cellulose, starch, and glycogen need to be split into glucose and other monosaccharides;
*   Proteins need to be split into amino acids;
*   Fats or lipids need to be split into fatty acids and glycerol.

The Gut

The digestive tract, alimentary canal or gut is a hollow tube stretching from the mouth to the anus. It is the organ system concerned with the treatment of foods.

At the mouth the large food molecules are taken into the gut - this is called ingestion. They must then be broken down into smaller ones by digestive enzymes - digestion, before they can be taken from the gut into the blood stream - absorption. The cells of the body can then use these small molecules - assimilation. The indigestible waste products are eliminated from the body by the act of egestion.

The 4 major functions of the gut are:

1. Transporting the food;

2. Processing the food physically by breaking it up (chewing), mixing, adding fluid etc.

3. Processing the food chemically by adding digestive enzymes to split large food molecules into smaller ones.

4. Absorbing these small molecules into the blood stream so the body can use them.

The regions of a typical mammals gut (for example a cat or dog) are shown.

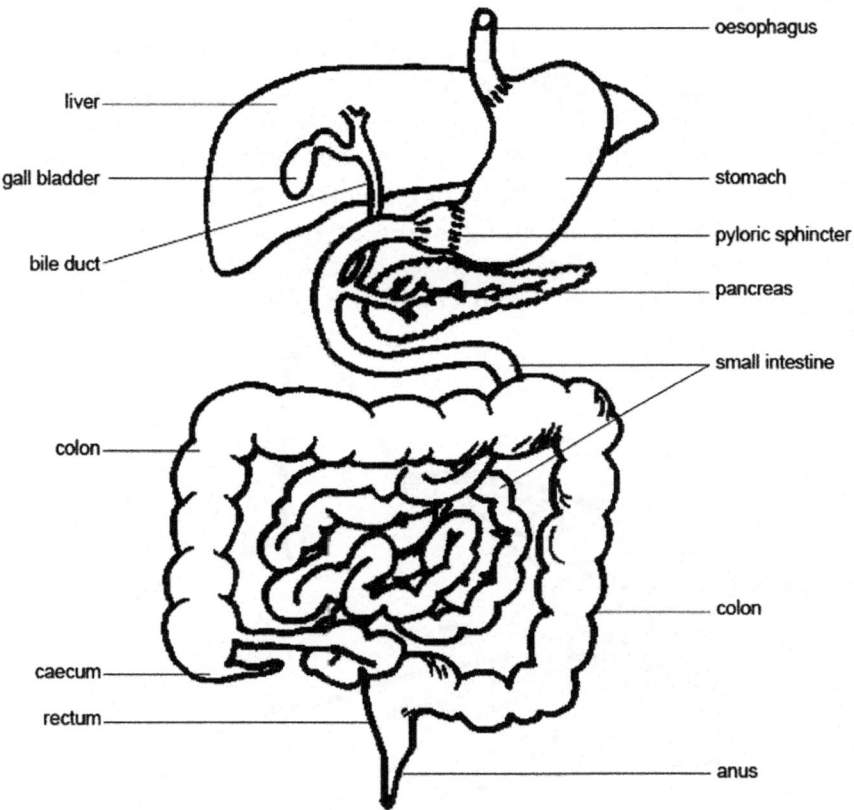

Figure 10: A typical mammalian gut

The food that enters the mouth passes to the oesophagus, then to the stomach, small intestine, caecum, large intestine, rectum and finally undigested material exits at the anus. The liver and pancreas produce secretions that aid digestion and the gall bladder stores bile.

Mouth

The mouth takes food into the body. The lips hold the food inside the mouth during chewing and allow the baby animal to suck on its mother's teat. In elephants the lips (and nose) have developed into the trunk which is the main food collecting tool. Some mammals, e.g. hamsters, have stretchy cheek pouches that they use to carry food or material to make their nests.

The sight or smell of food and its presence in the mouth stimulates the salivary glands to secrete saliva. There are four pairs of these glands in cats and dogs. The fluid they produce moistens and softens the food making it easier to swallow. It also contains the enzyme, salivary amylase, which starts the digestion of starch.

The tongue moves food around the mouth and rolls it into a ball for swallowing. Taste buds are located on the tongue and in dogs and cats it is covered with spiny projections used for grooming and lapping. The cow's tongue is prehensile and wraps around grass to graze it.

Swallowing is a complex reflex involving 25 different muscles. It pushes food into the oesophagus and at the same time a small flap of tissue called the epiglottis closes off the windpipe so food doesn't go 'down the wrong way' and choke the animal.

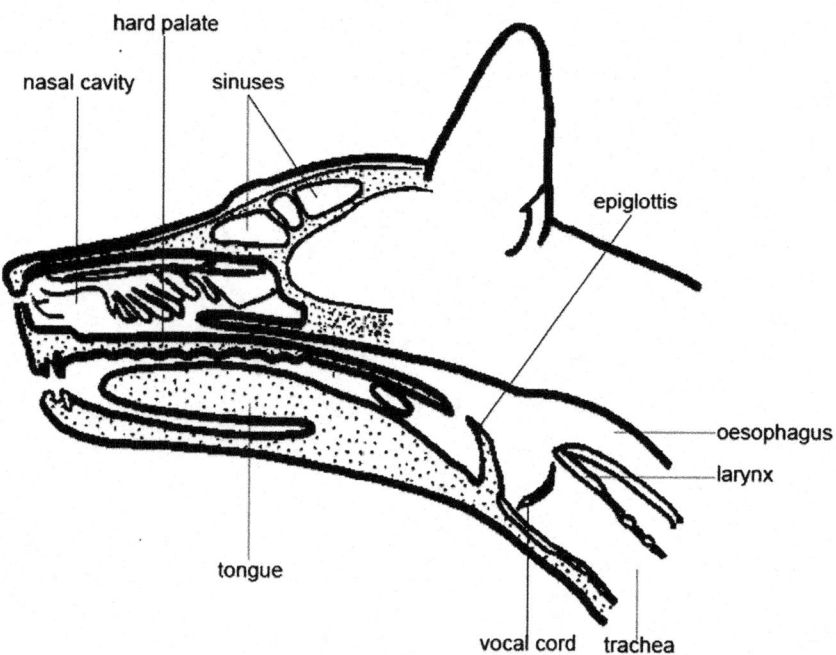

Figure 11: Section through the head of a dog

Teeth

Teeth seize, tear and grind food. They are inserted into sockets in the bone and consist of a crown above the gum and root below. The crown is covered with a layer of enamel, the hardest substance in the body. Below this is the dentine, a softer but tough and shock resistant material. At the centre of the tooth is a space filled with pulp which contains blood vessels and nerves. The tooth is cemented into the socket and in most teeth the tip of the root is quite narrow with a small opening for the blood vessels and nerves.

In teeth that grow continuously, like the incisors of rodents, the opening remains large and these teeth are called open rooted teeth. Mammals have 2 distinct sets of teeth. The first the milk teeth are replaced by the permanent teeth.

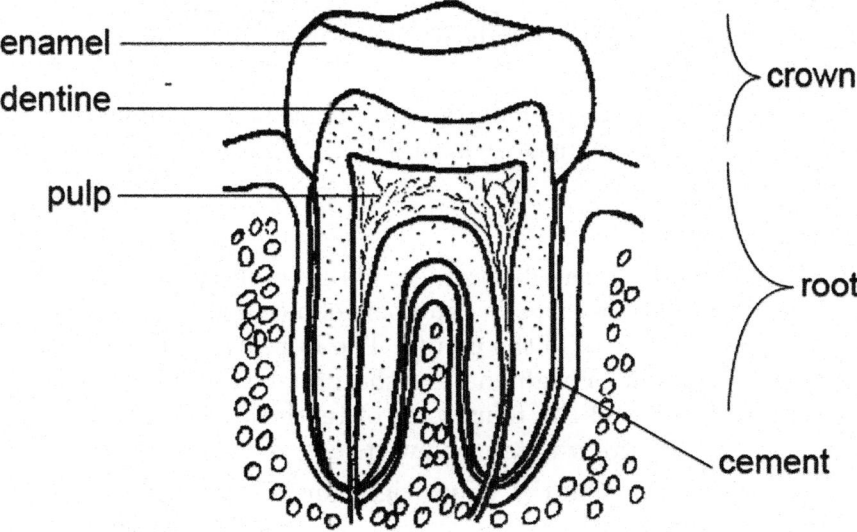

Figure 12: Structure of a tooth

Types of Teeth

All the teeth of fish and reptiles are similar but mammals usually have four different types of teeth.

The incisors are the chisel-shaped 'biting off' teeth at the front of the mouth. In rodents and rabbits the incisors never stop growing (open-rooted teeth). They must be worn or ground down continuously by gnawing. They have hard enamel on one surface only so they wear unevenly and maintain their sharp cutting edge.

The largest incisors in the animal kingdom are found in elephants, for tusks are actually giant incisors. Sloths have no incisors at all, and sheep have no incisors in the upper jaw. Instead there is a horny pad against which the bottom incisors cut. The canines or 'wolf-teeth' are long, cone-shaped teeth situated just behind the incisors. They are particularly well developed in the dog and cat families where they are used to hold, stab and kill the prey.

The tusks of boars and walruses are large canines while rodents and herbivores like sheep have no (or reduced) canines. In these animals the space where the canines would normally be is called the diastema. In rodents like the rat and beaver it allows the debris from gnawing to be expelled easily.

The cheek teeth or premolars and molars crush and grind the food. They are particularly well developed in herbivores where they have complex ridges that form broad grinding surfaces. These are created from alternating bands of hard enamel and softer dentine that wear at different rates.

In carnivores the premolars and molars slice against each other like scissors and are called carnassial teeth. They are used for shearing flesh and bone.

Dental Formula

The numbers of the different kinds of teeth can be expressed in a dental formula. This gives the numbers of incisors, canines, premolars and molars in one half of the mouth. The numbers of these four types of teeth in the left or right half of the upper jaw are written above a horizontal line and the four types of teeth in the right or left half of the lower jaw are written below it.

It indicates that in the upper right (or left) half of the jaw there are no incisors or canines (i.e. there is a diastema), three premolars and three molars. In the lower right (or left) half of the jaw are three incisors, one canine, three premolars and three molars.

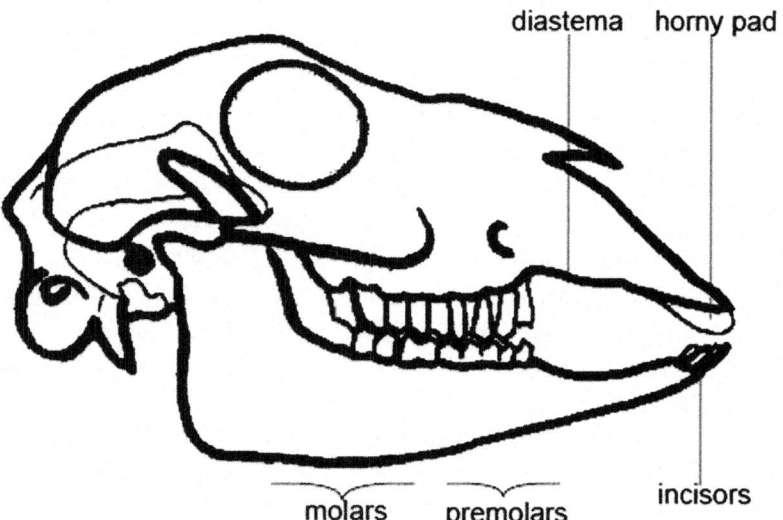

Figure 13: A sheep's skull

The formula indicates that in the right (or left) half of the upper jaw there are three incisors, one canine, four premolars and two molars. In the right (or left) half of the lower jaw there are three incisors, one canine, four premolars and three molars.

Oesophagus

The oesophagus transports food to the stomach. Food is moved along the oesophagus, as it is along the small and large intestines, by contraction of the smooth muscles in the walls that push the food along rather like toothpaste along a tube. This movement is called peristalsis.

Stomach

The stomach stores and mixes the food. Glands in the wall secrete gastric juice that contains enzymes to digest protein and fats as well as hydrochloric acid to make the contents very acidic. The walls of the stomach are very muscular and churn and mix the food with the gastric juice to form a watery mixture called chyme (pronounced kime). Rings of muscle called sphincters at the entrance and exit to the stomach control the movement of food into and out of it.

Small Intestine

Most of the breakdown of the large food molecules and absorption of the smaller molecules take place in the long and narrow small intestine. The total length varies but it is about 6.5 metres in humans, 21 metres in the horse, 40 metres in the ox and over 150 metres in the blue whale.

It is divided into 3 sections: the duodenum (after the stomach), jejunum and ileum. The duodenum receives 3 different secretions:

1) Bile from the liver;

2) Pancreatic juice from the pancreas and

3) Intestinal juice from glands in the intestinal wall.

These complete the digestion of starch, fats and protein. The products of digestion are absorbed into the blood and lymphatic system through the wall of the intestine, which is lined with tiny finger-like projections called villi that increase the surface area for more efficient absorption.

The Rumen

In ruminant herbivores like cows, sheep and antelopes the stomach is highly modified to act as a "fermentation vat". It is divided into four parts. The largest part is called the rumen. In the cow it occupies the entire left half of the abdominal cavity and can hold up to 270 litres.

The reticulum is much smaller and has a honeycomb of raised folds on its inner surface. In the camel the reticulum is further modified to store water. The next part is called the omasum with a folded inner surface. Camels have no omasum. The final compartment is called the abomasum. This is the 'true' stomach where muscular walls churn the food and gastric juice is secreted.

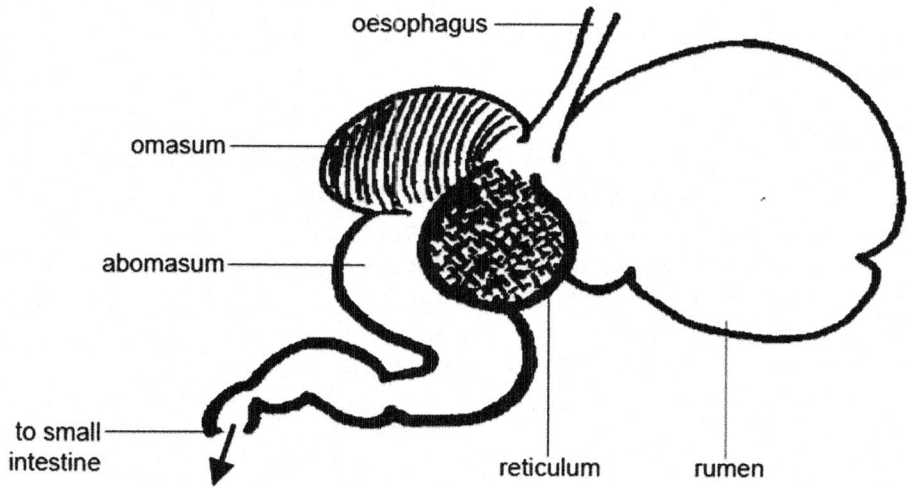

Figure 14: The rumen

Ruminants swallow the grass they graze almost without chewing and it passes down the oesophagus to the rumen and reticulum. Here liquid is added and the muscular walls churn the food. These chambers provide the main fermentation vat of the ruminant stomach. Here bacteria and single-celled animals start to act on the cellulose plant cell walls. These organisms break down the cellulose to smaller molecules that are absorbed to provide the cow or sheep with energy. In the process, the gases methane and carbon dioxide are produced. These cause the "burps" you may hear cows and sheep making.

Not only do the micro-organisms break down the cellulose but they also produce the vitamins E, B and K for use by the animal. Their digested bodies provide the ruminant with the majority of its protein requirements.

In the wild grazing is a dangerous activity as it exposes the herbivore to predators. They crop the grass as quickly as possible and then when the animal is in a safer place the food in the rumen can be regurgitated to be chewed at the animal's leisure. This is 'chewing the cud' or rumination. The finely ground food may be returned to the rumen for further work by the microorganisms or, if the particles are small enough, it will pass down a special groove in the wall of the oesophagus straight into the omasum. Here the contents are kneaded and water is absorbed before they pass to the abomasum. The abomasum acts as a "proper" stomach and gastric juice is secreted to digest the protein.

Large Intestine

The large intestine consists of the caecum, colon and rectum. The chyme from the small intestine that enters the colon consists mainly of water and undigested material such as cellulose (fibre or roughage). In omnivores like the pig and humans the main function of the colon is absorption of water to give solid faeces. Bacteria in this part of the gut produce vitamins B and K.

The caecum, which forms a dead-end pouch where the small intestine joins the large intestine, is small in pigs and humans and helps water absorption. However, in rabbits, rodents and horses, the caecum is very large and called the functional caecum. It is here that cellulose is digested by micro-organisms. The appendix, a narrow dead end tube at the end of the caecum, is particularly large in primates but seems to have no digestive function.

Functional Caecum

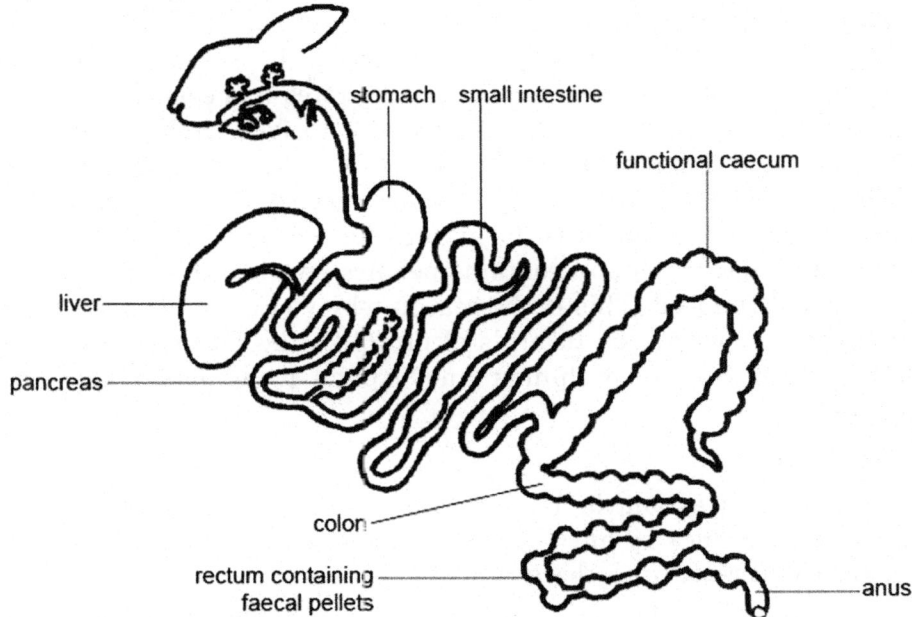

Figure 15: The gut of a rabbit

The caecum in the rabbit, rat and guinea pig is greatly enlarged to provide a "fermentation vat" for micro-organisms to break down the cellulose plant cell walls. This is called a functional caecum. In the horse both the caecum and the colon are enlarged. As in the rumen, the large cellulose molecules are broken down to smaller molecules that can be absorbed. However, the position of the functional caecum after the main areas of digestion and absorption, means it is potentially less effective than the rumen. This means that the small molecules that are produced there can not be absorbed by the gut but pass out in the faeces. The rabbit and rodents (and foals) solve this problem by eating their own faeces so that they pass through the gut a second time and the products of cellulose digestion can be absorbed in the small intestine. Rabbits produce two kinds of faeces. Softer night-time faeces are eaten directly from the anus and the harder pellets you are probably familiar with, that have passed through the gut twice.

The Gut of Birds

Birds' guts have important differences from mammals' guts. Most obviously, birds have a beak instead of teeth. Beaks are much lighter than teeth and are an adaptation for flight. Imagine a bird trying to take off and fly with a whole set of teeth in its head! At the base of the oesophagus birds have a bag-like structure called a crop. In many birds the crop stores food before it enters the stomach, while in pigeons and doves glands in the crop secretes a special fluid called crop-milk which parent birds regurgitate to feed their young. The stomach is also modified and consists of two compartments. The first is the true stomach with muscular walls and enzyme secreting glands. The second compartment is the gizzard. In seed eating birds this has very muscular walls and contains pebbles swallowed by the bird to help grind the food. This is the reason why you must always supply a caged bird with grit. In birds of prey like the falcon the walls of the gizzard are much thinner and expand to accommodate large meals.

Digestion

During digestion the large food molecules are broken down into smaller molecules by enzymes. The three most important groups of enzymes secreted into the gut are:

1. Amylases that split carbohydrates like starch and glycogen into monosaccharides like glucose.

2. Proteases that split proteins into amino acids.

3. Lipases that split lipids or fats into fatty acids and glycerol.

Glands produce various secretions which mix with the food as it passes along the gut.

These secretions include:

1. Saliva secreted into the mouth from several pairs of salivary glands. Saliva consists mainly of water but contains salts, mucous and salivary amylase. The function of saliva is to lubricate food as it is chewed and swallowed and salivary amylase begins the digestion of starch.

2. Gastric juice secreted into the stomach from glands in its walls. Gastric juice contains pepsin that breaks down protein and hydrochloric acid to produce the acidic conditions under which this enzyme works best. In baby animals rennin to digest milk is also produced in the stomach.

3. Bile produced by the liver. It is stored in the gall bladder and secreted into the duodenum via the bile duct. (Note that the horse, deer, parrot and rat have no gall bladder). Bile is not a digestive enzyme. Its function is to break up large globules of fat into smaller ones so the fat splitting enzymes can gain access the fat molecules.

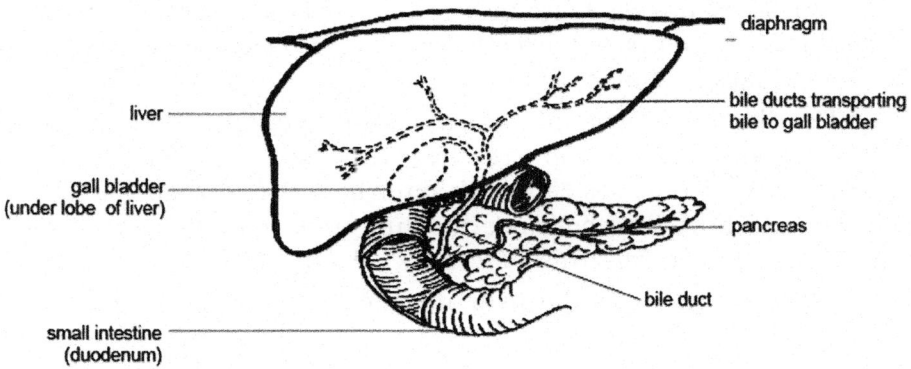

Figure 16: The liver, gall bladder and pancreas

Pancreatic Juice

The pancreas is a gland located near the beginning of the duodenum. In most animals it is large and easily seen but in rodents and rabbits it lies within the membrane linking the loops of the intestine (the mesentery) and is quite difficult to find. Pancreatic juice is produced in the pancreas. It flows into the duodenum and contains amylase for digesting starch, lipase for digesting fats and protease for digesting proteins.

Intestinal Juice

Intestinal juice is produced by glands in the lining of the small intestine. It contains enzymes for digesting disaccharides and proteins as well as mucus and salts to make the contents of the small intestine more alkaline so the enzymes can work.

Absorption

The small molecules produced by digestion are absorbed into the villi of the wall of the small intestine. The tiny finger-like projections of the villi increase the surface area for absorption. Glucose and amino acids pass directly through the wall into the blood stream by diffusion or active transport. Fatty acids and glycerol enter vessels of the lymphatic system (lacteals) that run up the centre of each villus.

The Liver

The liver is situated in the abdominal cavity adjacent to the diaphragm. It is the largest single organ of the body and has over 100 known functions. Its most important digestive functions are:

1. the production of bile to help the digestion of fats and
2. the control of blood sugar levels.

Glucose is absorbed into the capillaries of the villi of the intestine. The blood stream takes it directly to the liver via a blood vessel known as the hepatic portal vessel or vein.

The liver converts this glucose into glycogen which it stores. When glucose levels are low the liver can convert the glycogen back into glucose. It releases this back into the blood to keep the level of glucose constant. The hormone insulin, produced by special cells in the pancreas, controls this process.

Other functions of the liver include:

3. making vitamin A,
4. making the proteins that are found in the blood plasma (albumin, globulin and fibrinogen),
5. storing iron,
6. removing toxic substances like alcohol and poisons from the blood and converting them to safer substances,
7. producing heat to help maintain the temperature of the body.

Urinary System

Homeostasis

The cells of an animal can only remain healthy if the conditions are just right. The processes that take place in them are upset if the temperature is too high or too low, or if the fluid around or inside them is too acid or alkaline. Homeostasis is the name given to the processes that help keep the internal conditions constant even when external conditions change. The word means, "staying the same".

There are a number of organs in the body that play a part in maintaining homeostasis. For example, the skin helps keep the internal temperature of bird and mammals bodies within a narrow range even when the outside temperatures change; the lungs control the amount of carbon dioxide in the blood; the liver and pancreas work together to keep the amount of glucose in the blood within narrow limits, and the kidneys regulate the acidity and the concentration of water and salt in the blood.

Hormones are chemicals that carry messages around the body in the blood and are central to many of the homeostatic processes mentioned above.

Water In The Body

Water is essential for living things to survive because all the chemical reactions within a body take place in a solution of water. An animal's body consists of up to 80% water. The exact proportion depends on the type of animal, its age, sex, health and whether or not it has had sufficient to drink. Generally animals do not survive a loss of more than 15% of their body water.

In vertebrates almost 2/3rd of this water is in the cells (intracellular fluid). The rest is outside the cells (extracellular fluid) where it is found in the spaces around the cells (tissue fluid), as well as in the blood and lymph.

Maintaining Water Balance

Animals lose water through their skin and lungs, in the faeces and urine. These losses must be made up by water in food and drink and from the water that is a by-product of chemical reactions. If the animal does not manage to compensate for water loss the dissolved substances in the blood may become so concentrated they become lethal. To prevent this happening various mechanisms come into play as soon as the concentration of the blood increases. A part of the brain called the hypothalamus is in charge of these homeostatic processes. The most important is the feeling of thirst that is triggered by an increase in blood concentration. This stimulates an animal to find water and drink it.

The kidneys are also involved in maintaining water balance as various hormones instruct them to produce more concentrated urine and so retain some of the water that would otherwise be lost.

Desert Animals

Coping with water loss is a particular problem for animals that live in dry conditions. Some, like the camel, have developed great tolerance for dehydration. For example, under some conditions, camels can withstand the loss of one third of their body mass as water. They can also survive wide daily changes in temperature. This means they do not have to use large quantities of water in sweat to cool the body by evaporation.

Smaller animals are more able than large ones to avoid extremes of temperature or dry conditions by resting in sheltered more humid situations during the day and being active only at night.

The kangaroo rat is able to survive without access to any drinking water at all because it does not sweat and produces extremely concentrated urine. Water from its food and from chemical processes is sufficient to supply all its requirements.

Excretion

Animals need to excrete because they take in substances that are excess to the body's requirements and many of the chemical reactions in the body produce waste products. If these substances were not removed they would poison cells or slow down metabolism. All animals therefore have some means of getting rid of these wastes.

The major waste products in mammals are carbon dioxide that is removed by the lungs, and urea that is produced when excess amino acids (from proteins) are broken down. Urea is filtered from the blood by the kidneys.

The Kidneys And Urinary System

The kidneys in mammals are bean-shaped organs that lie in the abdominal cavity attached to the dorsal wall on either side of the spine. An artery from the dorsal aorta called the renal artery supplies blood to them and the renal vein drains them.

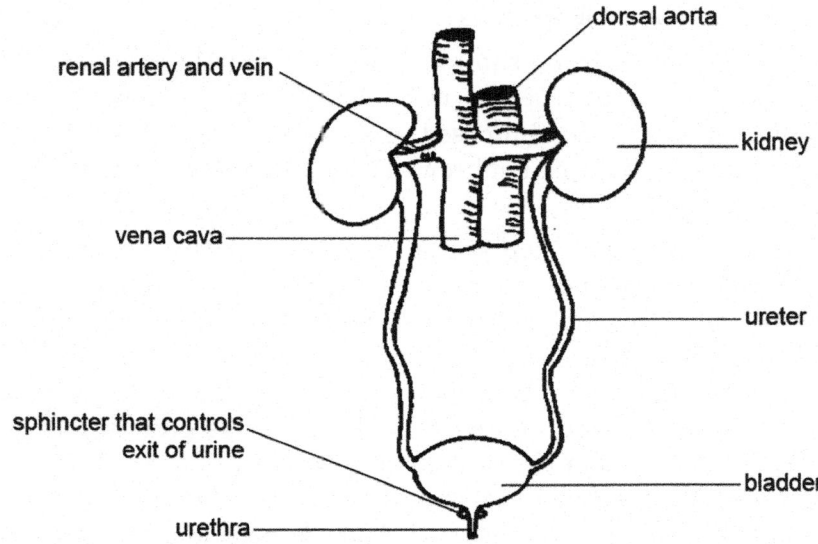

Figure 17: The urinary system

To the naked eye kidneys seem simple enough organs. They are covered by a fibrous coat or capsule and if cut in half lengthways (longitudinally) two distinct regions can be seen - an inner region or medulla and the outer cortex. A cavity within the kidney called the pelvis collects the urine and carries it to the ureter, which connects with the bladder where the urine is stored temporarily. Rings of muscle (sphincters) control the release of urine from the bladder and the urine leaves the body through the urethra.

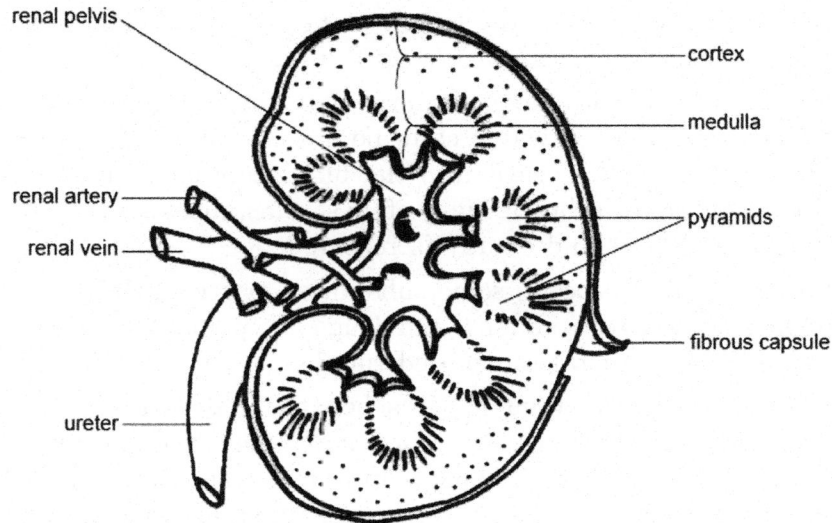

Figure 18: The dissected kidney

Kidney Tubules or Nephrons

It is only when you examine kidneys under the microscope that you find that their structure is not simple at all. The cortex and medulla are seen to be composed of masses of tiny tubes. These are called kidney tubules or nephrons. A human kidney consists of over a million of them.

At one end of each nephron, in the cortex of the kidney, is a cup shaped structure called the (Bowman's or renal) capsule. It surrounds a tuft of capillaries called the glomerulus that carries high-pressure blood. Together the glomerulus and capsule act as a blood-filtering device. The holes in the filter allow most of the contents of the blood through except the red and white cells and large protein molecules. The fluid flowing from the capsule into the rest of the kidney tubule is therefore very similar to blood plasma and contains many useful substances like water, glucose, salt and amino acids. It also contains waste products like urea.

Processes Occurring In The Nephron

After entering the glomerulus the filtered fluid flows along a coiled part of the tubule (the proximal convoluted tubule) to a looped portion (the Loop of Henle) and then to the collecting tube via a second length of coiled tube (the distal convoluted tubule). From the collecting ducts the urine flows into the renal pelvis and enters the ureter.

Note that the glomerulus, capsule and both coiled parts of the tubule are all situated in the cortex of the kidney while the loops of Henle and collecting ducts make up the medulla.

As the fluid flows along the proximal convoluted tubule useful substances like glucose, water, salts, potassium ions, calcium ions and amino acids are reabsorbed into the blood capillaries that form a network around the tubules. Many of these substances are transported by active transport and energy is required.

In a separate process, some substances, particularly potassium, ammonium and hydrogen ions, and drugs like penicillin, are actively secreted into the distal convoluted tubule.

By the time the fluid has reached the collecting ducts these processes of absorption and secretion have changed the fluid originally filtered into the Bowman's capsule into urine. The main function of the collecting ducts is then to remove more water from the urine if necessary. These processes are summarised.

Normal urine consists of water, in which waste products such as urea and salts such as sodium chloride are dissolved. Pigments from the breakdown of red blood cells give urine its yellow colour.

The Production Of Concentrated Urine

Because of the high pressure of the blood in the glomerulus and the large size of the pores in the glomerulus/capsule-filtering device, an enormous volume of fluid passes into the kidney tubules. If this fluid were left as it is, the animal's body would be drained dry in 30 minutes. In fact, as the fluid flows down the tubule, over 90% of the water in it is reabsorbed. The main part of this reabsorption takes place in the collecting tubes.

The amount of water removed from the collecting ducts is controlled by a hormone called antidiuretic hormone (ADH) produced by the pituitary gland, situated at the base of the brain. When the blood becomes more concentrated, as happens when an animal is deprived of water, ADH is secreted and causes more water to be absorbed from the collecting ducts so that concentrated urine is produced. When the animal has drunk plenty of water and the blood is dilute, no ADH is secreted and no or little water is absorbed from the collecting ducts, so dilute urine is produced. In this way the concentration of the blood is controlled precisely.

Water Balance In Fish and Marine Animals

Fresh Water Fish: Although the skin of fish is more or less waterproof, the gills are very porous. The body fluids of fish that live in fresh water have a higher concentration of dissolved substances than the water in which they swim. In other words the body fluids of fresh water fish are hypertonic to the water. Water therefore flows into the body by osmosis. To stop the body fluids being constantly diluted fresh water fish produce large quantities of dilute urine.

Marine Fish

Marine fish like the sharks and dogfish have body fluids that have the same concentration of dissolved substances as the water (isotonic) have little problem with water balance. However, marine bony fish like red cod, snapper and sole, have body fluids with a lower concentration of dissolved substances than seawater (they are hypotonic to seawater). This means that water tends to flow out of their bodies by osmosis. To make up this fluid loss they drink seawater and get rid of the excess salt by excreting it from the gills.

Marine Birds

Marine birds that eat marine fish take in large quantities of salt and some only have access to seawater for drinking. Bird's kidneys are unable to produce very concentrated urine, so they have developed a salt gland. This excretes a concentrated salt solution into the nose to get rid of the excess salt.

Diabetes and The Kidney

There are two types of diabetes. The most common is called sugar diabetes or diabetes mellitus and is common in cats and dogs especially if they are overweight. It is caused by the pancreas secreting insufficient insulin, the hormone that controls the amount of glucose in the blood. If insulin secretion is inadequate, the concentration of glucose in the blood increases. Any increase in the glucose in the blood automatically leads to an increase in glucose in the fluid filtered into the kidney tubule. Normally the kidney removes all the glucose filtered into it, but these high concentrations swamp this removal mechanism and urine containing glucose is produced. The main symptoms of this type of diabetes are the production of large amounts of dilute urine containing glucose, and excessive thirst.

The second type of diabetes is called diabetes insipidus. The name comes from the main symptom, which is the production of large amounts of very dilute and "tasteless" urine. It occurs when the pituitary gland produces insufficient ADH, the hormone that stimulates water reabsorption from the kidney tubule. When this hormone is lacking, water is not absorbed and large amounts of dilute urine are produced. Because so much water is lost in the urine, animals with this form of diabetes can die if deprived of water for only a day or so.

Other Functions of the Kidney

The excretion of urea from the body and the maintenance of water balance, as described above, are the main functions of the kidney. However, the kidneys have other roles in keeping conditions in the body stable i.e. in maintaining homeostasis. These include:

-   controlling the concentration of salt ions ($Na+$, $K+$, $Cl-$) in the blood by adjusting how much is excreted or retained;
-   maintaining the correct acidity of the blood. Excess acid is constantly being produced by the normal chemical reactions in the body and the kidney eliminates this.

Normal Urine

Normal urine consists of water (95%), urea, salts (mostly sodium chloride) and pigments (mostly from bile) that give it its characteristic colour.

Abnormal Ingredients of Urine

If the body is not working properly, small amounts of substances not normally present may be found in the urine or substances normally present may appear in abnormal amounts.

- The presence of glucose may indicate diabetes.

- Urine with red blood cells in it is called haematuria, and may indicate inflammaticn of the kidney or urinary tract, cancer or a blow to the kidneys.

- Sometimes free haemoglobin is found in the urine. This indicates that the red blood cells in the blood have haemolysed (the membrane has broken down) and the haemoglobin has passed into the kidney tubules.

- The presence of white blood cells in the urine indicates there is an infection in the kidney or urinary tract.

- Protein molecules are usually too large to pass into the kidney tubule so no or only small amounts of proteins like albumin is normally found in urine. Large quantities of albumin indicate that the kidney tubules have been injured or the kidney has become diseased. High blood pressure also pushes proteins from the blood into the tubules.

- Casts are tiny cylinders of material that have been shed from the lining of the tubules and flushed out into the urine.

- Mucus is not usually found in the urine of healthy animals but is a normal constituent of horses' urine, giving it a characteristic cloudy appearance.

Tests can be carried out to identify any abnormal ingredients of urine. These tests are normally done by "stix", which are small plastic strips with absorbent ends impregnated with various chemicals. A colour change occurs in the presence of an abnormal ingredient.

Excretion in Birds

Birds' high body temperature and level of activity means that they need to conserve water. Birds therefore do not have a bladder and instead of excreting urea, which needs to be dissolved in large amounts of water, birds produce uric acid that can be discharged as a thick paste along with the feces. This is the white chalky part of the bird droppings that land on you or your car.

Nervous System

Coordination

Animals must be able to sense and respond to the environment in which they live if they are to survive. They need to be able to sense the temperature of their surroundings, for example, so they can avoid the hot sun. They must also be able to identify food and escape predators.

The various systems and organs in the body must also be linked so they work together. For example, once a predator has identified suitable prey it has to catch it. This involves coordinating the contraction of the muscle so the predator can run, there must then be an increased blood supply to the muscles to provide them with oxygen and nutrients. At the same time the respiration rate must increase to supply the oxygen and remove the carbon dioxide produced as a result of this increased activity. Once the prey has been caught and eaten, the digestive system must be activated to digest it.

The adjustment of an animal's response to changes in the environment and the complex linking of the various processes in the body that this response involves are called co-ordination. Two systems are involved in co-ordination in animals. These are the nervous and endocrine systems. The first operates via electrical impulses along nerve fibres and the second by releasing special chemicals or hormones into the bloodstream from glands.

Functions of the Nervous System

The nervous system has three basic functions:

1. *Sensory Function:* to sense changes (known as stimuli) both outside and within the body. For example the eyes sense changes in light and the ear responds to sound waves. Inside the body, tretch receptors in the stomach indicate when it is full and chemical receptors in the blood vessels monitor the acidity of the blood.

2. *Integrative Function:* processing the information received from the sense organs. The impulses from these organs are analysed and stored as memory. The many different impulses from

different sources are sorted, synchronised and co-ordinated and the appropriate response initiated. The power to integrate, remember and apply experience gives higher animals much of their superiority.

3. *Motor function:* The third function is the response to the stimuli that causes muscles to contract or glands to secrete.

All nervous tissue is made up of nerve cells or neurons. These transmit high-speed signals called nerve impulses. Nerve impulses can be thought of as being similar to an electric current.

The Neuron

Neurons are cells that have been adapted to carry nerve impulses. A typical neuron has a cell body containing a nucleus, one or more branching filaments called dendrites which conduct nerve impulses towards the cell body and one long fibre, an axon, that carries the impulses away from it. Many axons have a sheath of fatty material called myelin surrounding them. This speeds up the rate at which the nerve impulses travel along the nerve.

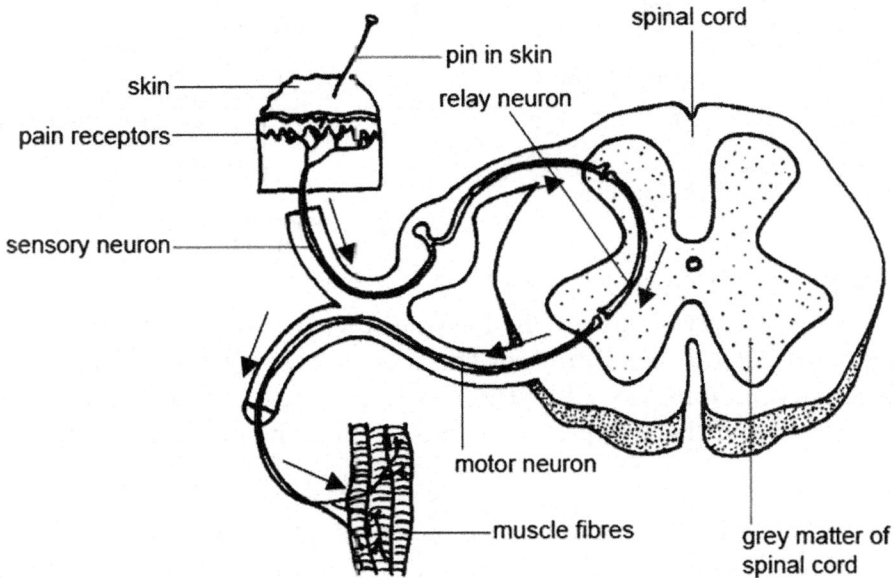

Figure 19: The relationship between sensory, relay and motor neurons

The cell body of neurons is usually located in the brain or spinal cord while the axon extends the whole distance to the organ that it supplies. The neuron carrying impulses from the spinal cord to the hind leg or tail of a horse, for example, can be several feet long. A nerve is a bundle of axons. A sensory neuron is a nerve cell that transmits impulses from a sense receptor such as those in the eye or ear to the brain or spinal cord. A motor neuron is a nerve cell that transmits impulses from the brain or spinal cord to a muscle or gland. A relay neuron connects sensory and motor neurons and is found in the brain or spinal cord.

Connections between Neurons

The connection between adjacent neurons is called a synapse. The two nerve cells do not actually touch here for there is a microscopic space between them. The electrical impulse in the neurone before the synapse stimulates the production of chemicals called neurotransmitters (such as acetylcholine), which are secreted into the gap.

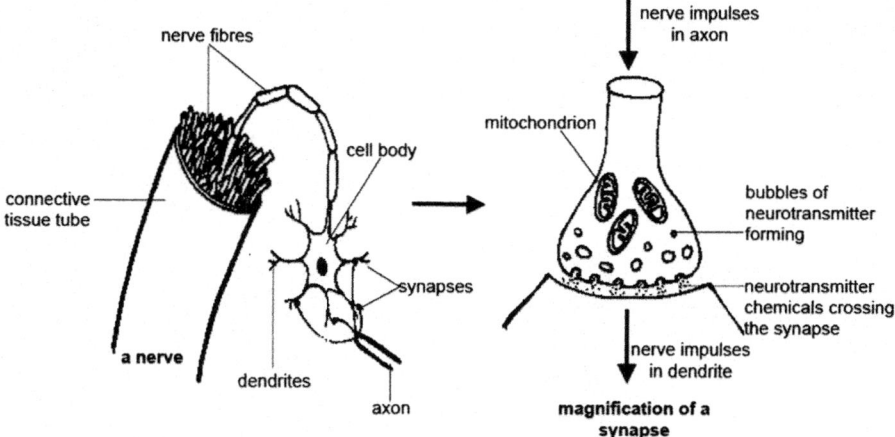

Figure 20: A nerve and magnification of a synapse

The neurotransmitter chemicals diffuse across the gap and when they contact the membrane of the next nerve cell they stimulate a new nervous impulse. After the impulse has passed the chemical is destroyed and the synapse is ready to receive the next nerve impulse.

Reflexes

A reflex is a rapid automatic response to a stimulus. When you accidentally touch a hot object and automatically jerk your hand away, this is a reflex action. It happens without you having to think about it. Animals automatically blink when an object approaches the eye and cats twist their bodies in the air when falling so they land on their paws. Swallowing, sneezing, and the constriction of the pupil of the eye in bright light are also all reflex actions.

The path taken by the nerve impulses in a reflex is called a reflex arc. Most reflex arcs involve only three neurons. The stimulus (a pin in the paw) stimulates the pain receptors of the skin, which initiate an impulse in a sensory neuron.

This travels to the spinal cord where it passes, by means of a synapse, to a connecting neuron called the relay neuron situated in the spinal cord. The relay neuron in turn makes a synapse with one or more motor neurons that transmit the impulse to the muscles of the limb causing them to contract and remove the paw from the sharp object. Reflexes do not require involvement of the brain although you are aware of what is happening and can, in some instances, prevent them happening. Animals are born with their reflexes. You can think of them as being wired in.

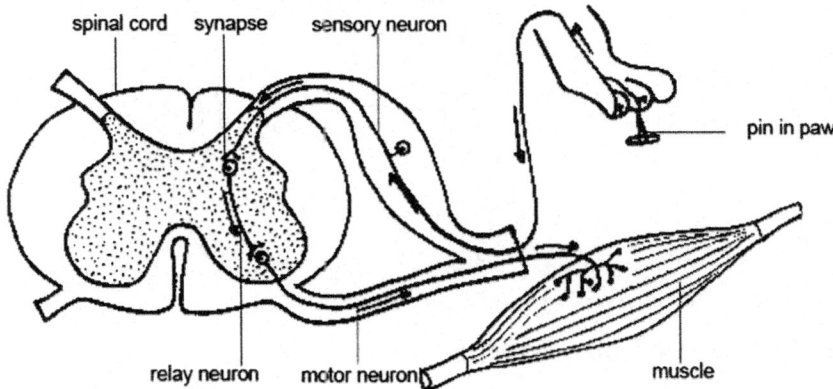

Figure 21: A reflex arc

Conditioned Reflexes

In most reflexes the stimulus and response are related. For example the presence of food in the mouth causes the salivary glands to release saliva. However, it is possible to train animals (and humans) to respond to different and often quite irrelevant stimuli. This is called a conditioned reflex.

A Russian biologist called Pavlov carried out the classic experiment to demonstrate such a reflex when he conditioned dogs to salivate at the sound of a bell ringing. Almost every pet owner can identify reflexes they have conditioned in their animals. Perhaps you have trained your cat to associate food with the opening of the fridge door or accustomed your dog to the routines you go through before taking them for a walk.

Parts of the Nervous System

When we describe the nervous system of vertebrates we usually divide it into two parts.

1. The central nervous system (CNS) which consists of the brain and spinal cord.

2. The peripheral nervous system (PNS) which consists of the nerves that connect to the brain and spinal cord (cranial and spinal nerves) as well as the autonomic (or involuntary) nervous system.

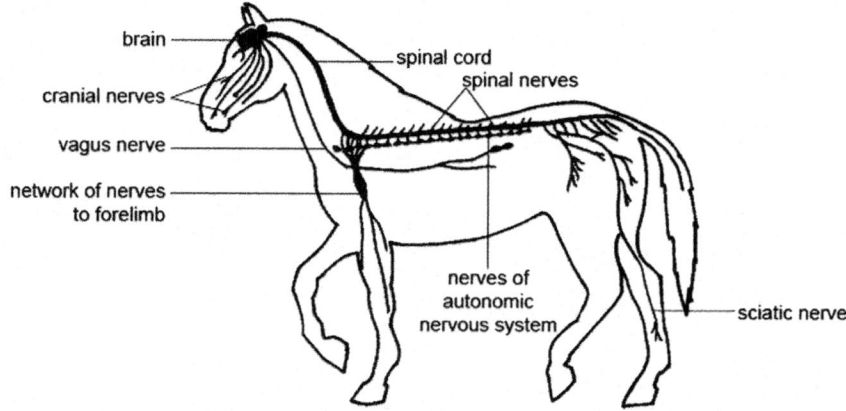

Figure 22: The nervous system of a horse

The Central Nervous System

The central nervous system consists of the brain and spinal cord. It acts as a kind of 'telephone exchange' where a vast number of cross connections are made.

When you look at the brain or spinal cord some regions appear creamy white (white matter) and others appear grey (grey matter). White matter consists of masses of nerve axons and the grey matter consists of the nerve cells. In the brain the grey matter is on the outside and in the spinal cord it is on the inside.

The Brain

The major part of the brain lies protected within the sturdy "box" of skull called the cranium. Surrounding the fragile brain tissue (and spinal cord) are protective membranes called the meninges, and a crystal-clear fluid called cerebrospinal fluid, which protects and nourishes the brain tissue. This fluid also fills four cavities or ventricles that lie within the brain.

Brain tissue is extremely active and, even when an animal is resting, it uses up to 20% of the oxygen taken into the body by the lungs. The carotid artery, a branch off the dorsal aorta, supplies it with the oxygen and nutrients it requires. Brain damage occurs if brain tissue is deprived of oxygen for only 4-8 minutes.

The brain consists of three major regions:

1. the fore brain which includes the cerebral hemispheres, hypothalamus and pituitary gland;

2. the hind brain or brain stem, contains the medulla oblongata and pons and

3. the cerebellum or "little brain".

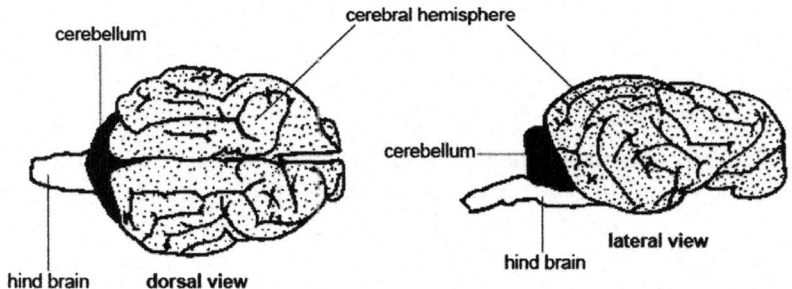

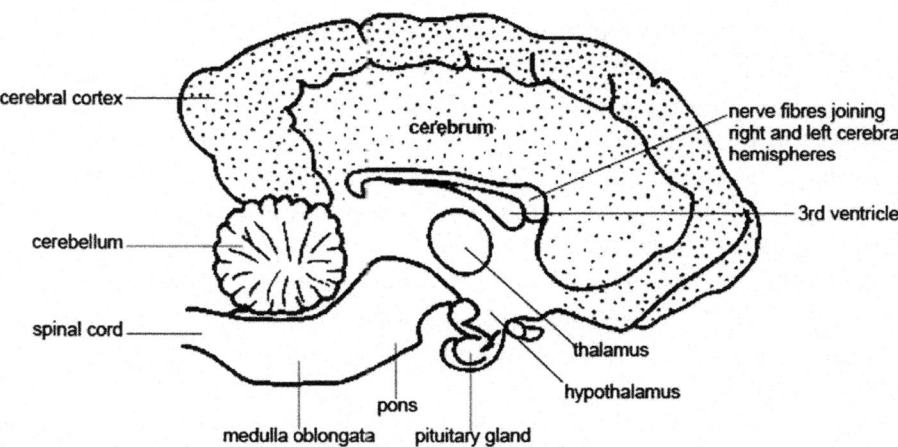

Figure 23: Longitudinal section through the brain of a dog

Mapping the Brain

In humans and some animals the functions of the different regions of the cerebral cortex have been mapped.

The Forebrain

The cerebral hemispheres are the masses of brain tissue that sit on the top of the brain. The surface is folded into ridges and furrows called sulci (singular sulcus). They make this part of the brain look rather like a very large walnut kernel. The two hemispheres are separated by a deep groove although they are connected internally by a thick bundle of nerve fibres. The outer layer of each hemisphere is called the cerebral cortex and this is where the main functions of the cerebral hemispheres are carried out.

The cerebral cortex is large and convoluted in mammals compared to other vertebrates and largest of all in humans because this is where the so-called "higher centres" concerned with memory, learning, reasoning and intelligence are situated.

Nerves from the eyes, ears, nose and skin bring sensory impulses to the cortex where they are interpreted. Appropriate voluntary movements are initiated here in the light of the memories of past events.

Different regions of the cortex are responsible for particular sensory and motor functions, e.g. vision, hearing, taste, smell, or moving the fore-limbs, hind-limbs or tail. For example, when a dog sniffs a scent, sensory impulses from the organ of smell in the nose pass via the olfactory (smelling) nerve to the olfactory centres of the cerebral hemispheres where the impulses are interpreted and co-ordinated.

In humans and some animals the functions of the different regions of the cerebral cortex have been mapped.

The hypothalamus is situated at the base of the brain and is connected by a "stalk" to the pituitary gland, the "master" hormone-producing gland. The hypothalamus can be thought of as the bridge between the nervous and endocrine (hormone producing) systems. It produces some of the hormones that are released from the pituitary gland and controls the release of others from it.

It is also an important centre for controlling the internal environment of the animal and therefore maintaining homeostasis. For example, it helps regulate the movement of food through the gut and the temperature, blood pressure and concentration of the blood. It is also responsible for the feeling of being hungry or thirsty and it controls sleep patterns and sex drive.

The Hindbrain

The medulla oblongata is at the base of the brain and is a continuation of the spinal cord. It carries all signals between the spinal cord and the brain and contains centres that control vital body functions like the basic rhythm of breathing, the rate of the heartbeat and the activities of the gut. The medulla oblongata also co-ordinates swallowing, vomiting, coughing and sneezing.

The Cerebellum

The cerebellum (little brain) looks rather like a smaller version of the cerebral hemispheres attached to the back of the brain. It receives impulses from the organ of balance (vestibular organ) in the inner ear and from stretch receptors in the muscles and tendons. By co-ordinating these it regulates muscle contraction during walking and running and helps maintain the posture and balance of the animal. When the cerebellum malfunctions it causes a tremor and uncoordinated movement.

The Spinal Cord

The spinal cord is a cable of nerve tissue that passes down the channel in the vertebrae from the hindbrain to the end of the tail. It becomes progressively smaller as paired spinal nerves pass out of the cord to parts of the body. Protective membranes or meninges cover the cord and these enclose cerebral spinal fluid.

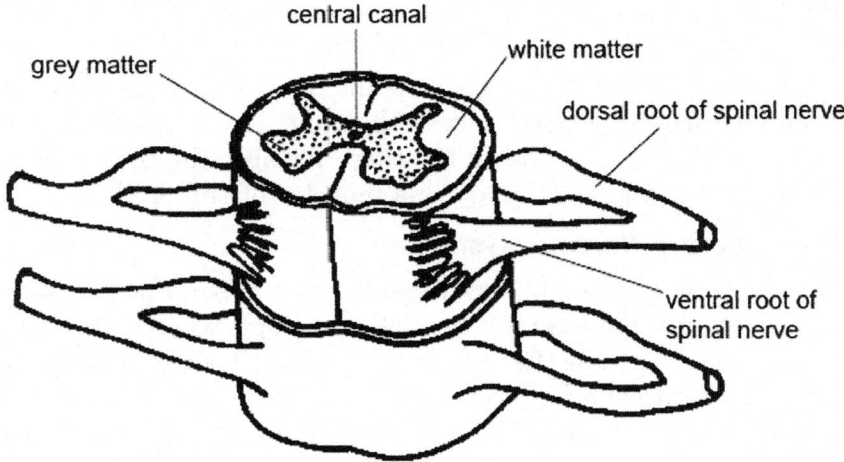

Figure 24: The spinal cord

If you cut across the spinal cord you can see that it consists of white matter on the outside and grey matter in the shape of an H or butterfly on the inside.

The Peripheral Nervous System

The peripheral nervous system consists of nerves that are connected to the brain (cranial nerves), and nerves that are connected to the spinal cord (spinal nerves). The autonomic nervous system is also part of the peripheral nervous system.

Cranial Nerves

There are twelve pairs of cranial nerves that come from the brain. Each passes through a hole in the cranium (brain case). The most important of these are the olfactory, optic, acoustic and vagus nerves.

The olfactory nerves - (smell) carry impulses from the olfactory organ of the nose to the brain.

The optic nerves - (sight) carry impulses from the retina of the eye to the brain.

The auditory (acoustic) nerves - (hearing) carry impulses from the cochlear of the inner ear to the brain.

The vagus nerve - controls the muscles that bring about swallowing. It also controls the muscles of the heart, airways, lungs, stomach and intestines.

Spinal Nerves

Spinal nerves connect the spinal cord to sense organs, muscles and glands in the body. Pairs of spinal nerves leave the spinal cord and emerge between each pair of adjacent vertebrae.

The sciatic nerve is the largest spinal nerve in the body. It leaves the spinal cord as several nerves that join to form a flat band of nervous tissue. It passes down the thigh towards the hind leg where it gives off branches to the various muscles of this limb.

The Autonomic Nervous System

The autonomic nervous system controls internal body functions that are not under conscious control. For example when a prey animal is chased by a predator the autonomic nervous system automatically increases the rate of breathing and the heartbeat. It dilates the blood vessels that carry blood to the muscles, releases glucose from the liver, and makes other adjustments to provide for the sudden increase in activity. When the animal has escaped and is safe once again the nervous system slows down all these processes and resumes all the normal body activities like the digestion of food.

The nerves of the autonomic nervous system originate in the spinal cord and pass out between the vertebrae to serve the various organs. There are two main parts to the autonomic nervous system — the sympathetic system and the parasympathetic system.

The sympathetic system stimulates the "flight, fright, fight" response that allows an animal to face up to an attacker or make a rapid departure. It increases the heart and respiratory rates, as well as the amount of blood flowing to the skeletal muscles while blood flow to less critical regions like the gut and skin is reduced. It also causes the pupils of the eyes to dilate. Note that the effects of the sympathetic system are similar to the effects of the hormone adrenaline.

The parasympathetic system does the opposite to the sympathetic system. It maintains the normal functions of the relaxed body. These are sometimes known as the "housekeeping" functions. It promotes effective digestion, stimulates defaecation and urination and maintains a regular heartbeat and rate of breathing.

The Senses

The Sense Organs

Sense organs allow animals to sense changes in the environment around them and in their bodies so that they can respond appropriately. They enable animals to avoid hostile environments, sense the presence of predators and find food.

Animals can sense a wide range of stimuli that includes, touch, pressure, pain, temperature, chemicals, light, sound, movement and position of the body. Some animals can sense electric and magnetic fields. All sense organs respond to stimuli by producing nerve impulses that travel to the brain via a sensory nerve. The impulses are then processed and interpreted in the brain as pain, sight, sound, taste etc.

The senses are often divided into two groups:

1. The general senses of touch, pressure, pain and temperature that are distributed fairly evenly through the skin. Some are found in muscles and within joints.
2. The special senses which include the senses of smell, taste, sight, hearing and balance. The special sense organs may be quite complex in structure.

Touch and Pressure

Within the dermis of the skin are numerous modified nerve endings that are sensitive to touch and pressure. The roots of hairs may also be well supplied with sensory receptors that inform the animal that it is in contact with an object. Whiskers are specially modified hairs.

Pain

Receptors that sense pain are found in almost every tissue of the body. They tell the animal that tissues are dangerously hot, cold, compressed or stretched or that there is not enough blood flowing in them. The animal may then be able to respond and protect itself from further damage

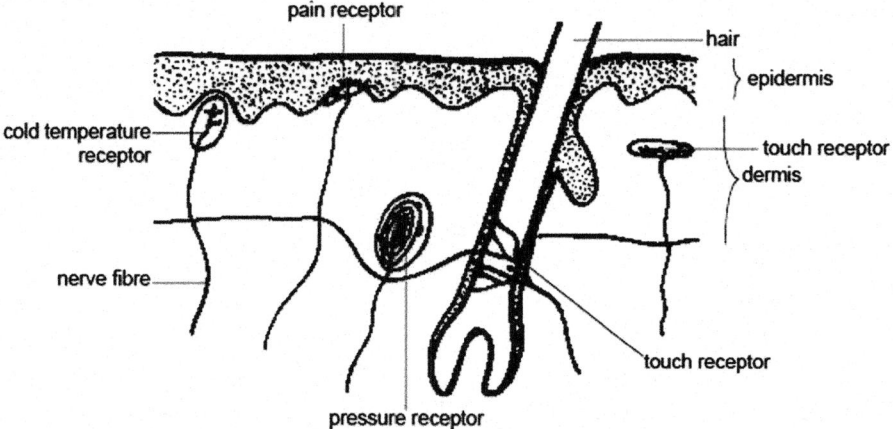

Figure 25: The general senses in the skin

Temperature

Nerve endings in the skin respond to hot and cold stimuli.

Awareness of Limb Position

There are sense organs in the muscles, tendons and joints that send continuous impulses to the brain that tell it where each limb is. This information allows the animal to place its limbs accurately and know their exact position without having to watch them.

Smell

Animals use the sense of smell to locate food, mark territory, identify their own offspring and the presence and sexual condition of a potential mate. The organ of smell (olfactory organ) is located in the nose and responds to chemicals in the air. It consists of modified nerve cells that have several tiny hairs on the surface. These emerge from the epithelium on the roof of the nose cavity into the mucus that lines it. As the animal breathes, chemicals in the air dissolve in the mucus. When the sense cell responds to a particular molecule, it fires an impulse that travels along the olfactory nerve to the brain where it is interpreted as an odour.

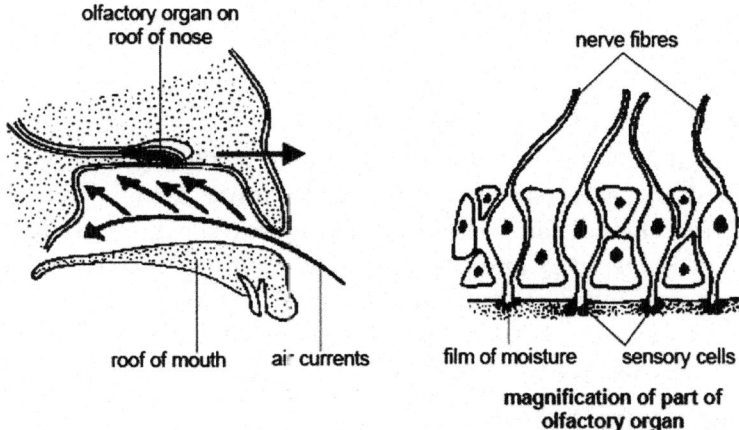

Figure 26: The olfactory organ - the sense of smell

The olfactory sense in humans is rudimentary compared to that of many animals. Carnivores that hunt have a very highly developed sensitivity to scents. For example a polar bear can smell out a dead seal 20 km away and a bloodhound can distinguish between the trails of different people although it may sometimes be confused by the criss-crossing trail of identical twins.

Snakes and lizards detect odours by means of Jacobson's organ. This is situated on the roof of the mouth and consists of pits containing sensory cells. When snakes flick out their forked tongues they are smelling the air by carrying the molecules in it to the Jacobson's organ.

Taste

The sense of taste allows animals to detect and identify dissolved chemicals. In reptiles, birds, and mammals the taste receptors (taste buds) are found mainly to the upper surface of the tongue. They consist of pits containing sensory cells arranged rather like the segments of an orange. Each receptor cell has a tiny "hair" that projects into the saliva to sense the chemicals dissolved in it.

The sense of taste is quite restricted. Humans can only distinguish four different tastes (sweet, sour, bitter and salt) and what we normally think of as taste is mainly the sense of smell. Food is quite tasteless when the nose is blocked and cats often refuse to eat when this happens.

Sight

The eyes are the organs of sight. They consist of spherical eyeballs situated in deep depressions in the skull called the orbits. They are attached to the wall of the orbit by six muscles, which move the eyeball. Upper and lower eyelids cover the eyes during sleep and protect them from foreign objects or too much light, and spread the tears over their surface.

The nictitating membrane or haw is a transparent sheet that moves sideways across the eye from the inner corner, cleansing and moistening the cornea without shutting out the light. It is found in birds, crocodiles, frogs and fish as well as marsupials like the kangaroo. It is rare in mammals but can be seen in cats and dogs by gently opening the eye when it is asleep. Eyelashes also protect the eyes from the sun and foreign objects.

Structure of the Eye

Lining the eyelids and covering the front of the eyeball is a thin epithelium called the conjunctiva. Conjunctivitis is inflammation of this membrane. Tear glands that open just under the top eyelid secrete a salty solution that keeps the exposed part of the eye moist, washes away dust and contains an enzyme that destroys bacteria.

The wall of the eyeball is composed of three layers. From the outside these are the sclera, the choroid and the retina.

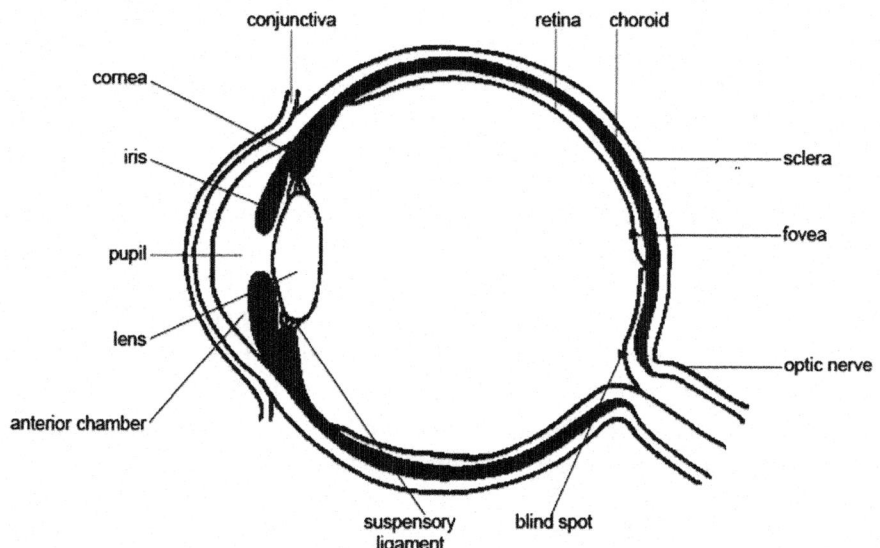

Figure 27: The structure of the eye

The sclera is a tough fibrous layer that protects the eyeball and gives it rigidity. At the front of the eye the sclera is visible as the "white" of the eye, which is modified as the transparent cornea through which the light rays have to pass to enter the eye. The cornea helps focus light that enters the eye.

The choroid lies beneath the sclera. It contains a network of blood vessels that supply the eye with oxygen and nutrients. Its inner surface is highly pigmented and absorbs stray light rays. In nocturnal animals like the cat and possum this highly pigmented layer reflects light as a means of conserving light. This is what makes them shine when caught in car headlights.

At the front of the eye the choroid becomes the iris. This is the coloured part of the eye that controls the amount of light entering the pupil of the eye. In dim light the pupil is wide open so as much light as possible enters while in bright light the pupils contract to protect the retina from damage by excess light.

The pupil in most animals is circular but in many nocturnal animals it is a slit that can close completely. This helps protect the extra-sensitive light sensing tissues of animals like the cat and possum from bright sunlight.

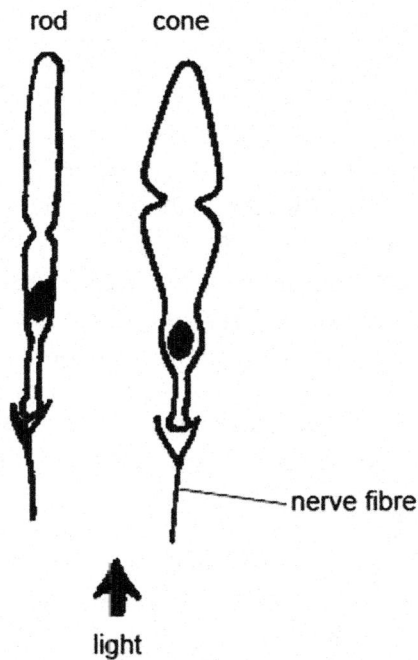

Figure 28: A rod and cone from the retina

The inner layer lining the inside of the eye is the retina. This contains the light sensing cells called rods and cones.

The rod cells are long and fat and are sensitive to dim light but cannot detect colour. They contain large amounts of a pigment that changes when exposed to light. This pigment comes from vitamin A found in carrots etc. A deficiency of this vitamin causes night blindness. So your mother was right when she told you to eat your carrots as they would help you see in the dark.

The cone cells provide colour vision and allow animals to see details. Most are found in the centre of the retina and they are most densely concentrated in a small area called the fovea. This is the area of sharpest vision, where the words you are reading at this moment are focussed on your retina.

The nerve fibres from the cells of the retina join and leave the eye via the optic nerve. There are no rods or cones here and it is a blind spot. The optic nerve passes through the back of the orbit and enters the brain.

The lens is situated just behind the pupil and the iris. It is a crystalline structure with no blood vessels and is held in position by a ligament.

This is attached to a muscle, which changes the shape of the lens so both near and distant objects can be focussed by the eye. This ability to change the focus of the lens is called accommodation. In many mammals the muscles that bring about accommodation are poorly developed, Rats, cows and dogs, for example, are thought to be unable to focus clearly on near objects.

In old age and certain diseases the lens may become cloudy resulting in blurred vision. This is called a cataract. Within the eyeball are two cavities, the anterior and posterior chambers, separated by the lens. They contain fluids the aqueous and vitreous humours respectively, that maintain the shape of the eyeball and help press the retina firmly against the choroid so clear images are seen.

How the Eye Sees

Eyes work quite like a camera. Light rays from an object enter the eye and are focused on the retina (the "film") at the back of the eye. The cornea, the lens and the fluid within the eye all help to focus the light. They do this by bending the light rays so that light from the object falls on the retina. This bending of light is called refraction. The light stimulates the light sensitive cells of the retina and nerve impulses are produced that pass down the optic nerve to the brain.

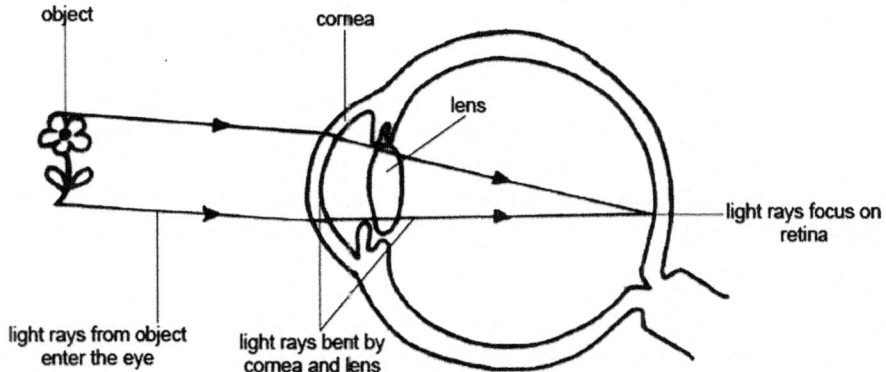

Figure 29: How the light travels from the object to the retina of the eye

Colour Vision in Animals

As mentioned before, the retina has two different kinds of cells that are stimulated by light - *'rods'* and *'cones'*. In humans and higher primates like baboons and gorillas the rods function in dim light and do not perceive colour, while the cones are stimulated by bright light and perceive details and colour.

Other mammals have very few cones in their retinas and it is believed that they see no or only a limited range of colour. It is, of course, difficult to find out exactly what animals do see. It is thought that deer, rats, and rabbits and nocturnal animals like the cat are colour-blind, and dogs probably see green and blue. Some fish and most birds seem to have better colour vision than humans and they use colour, often very vivid ones, for recognising each other as well as for courtship and protection.

Binocular Vision

Animals like cats that hunt have eyes placed on the front of the head in such a way that both eyes see the same wide area but from slightly different angles. This is called binocular vision. Its main advantage is that it enables the animals to estimate the distance to the prey so they can chase it and pounce accurately.

In contrast plant-eating prey animals like the rabbit and deer need to have a wide panoramic view so they can see predators approaching. They therefore have eyes placed on the side of the head, each with its own field of vision. They have only a very small area of binocular vision in front of the head but are extremely sensitive to movement.

Hearing

Animals use the sense of hearing for many different purposes. It is used to sense danger and enemies, to detect prey, to identify prospective mates and to communicate within social groups. Some animals (e.g. most bats and dolphins) use sound to "see" by echolocation. By sending out a cry and interpreting the echo, they sense obstacles or potential prey.

Structure of the Ear

Most of the ear, the organ of hearing, is hidden from view within the bony skull. It consists of three main regions: the outer ear, the middle ear and the inner ear.

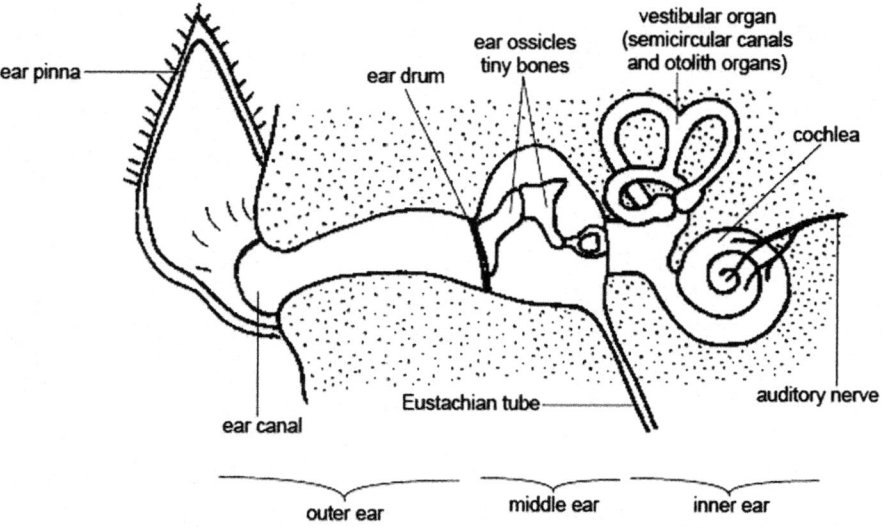

Figure 30: The ear

The outer ear consists of an ear canal leading inwards to a thin membrane known as the eardrumour tympanic membrane that stretches across the canal. Many animals have an external ear flap or pinna to collect and funnel the sound into the ear canal. The pinnae (plural of pinna) usually face forwards on the head but many animals can swivel them towards the source of the sound.

In dogs the ear canal is long and bent and often traps wax or provides an ideal habitat for mites, yeast and bacteria.

The middle ear consists of a cavity in the skull that is connected to the pharynx (throat) by a long narrow tube called the Eustachian tube. This links the middle ear to the outside air so that the air pressure on both sides of the eardrum can be kept the same. Everyone knows the uncomfortable feeling (and affected hearing) that occurs when you drive down a steep hill and the unequal air pressures on the two sides of the eardrum cause it to distort. The discomfort is relieved when you swallow because the Eustachian tubes open and the pressure on either side equalises.

Within the cavity of the middle ear are three of the smallest bones in the body, the auditory ossicles. They are known as the hammer, the anvil and the stirrup because of their resemblance to the shape of these objects. These tiny bones articulate (move against) each other and transfer the vibrations of the eardrum to the membrane covering the opening to the inner ear.

The inner ear is a complicated series of fluid-filled tubes imbedded in the bone of the skull. It consists of two main parts. These are the cochlea where sound waves are converted to nerve impulses and the vestibular organ that is associated with the sense of balance and has no role in hearing.

The cochlea looks rather like a coiled up snail shell. Within it there are specialised cells with fine hairs on their surface that respond to the movement of the fluid within the cochlea by producing nervous impulses that travel to the brain along the auditory nerve.

How The Ear Hears

Sound waves can be thought of as vibrations in the air. They are collected by the ear pinna and pass down the ear canal where they cause the eardrum to vibrate. (An interesting fact is that when you are listening to someone speaking your eardrum vibrates at exactly the same rate as the vocal cords of the person speaking to you).

The vibration of the eardrum sets the three tiny bones in the middle ear moving against each other so that the vibration is transferred to the membrane covering the opening to the inner ear. As well as transferring the vibration, the tiny ear bones also amplify it. The three tiny bones are called the stirrup, anvil, and hammer. They were called such of their form. In the human ear this amplification is about 20 times while in desert-dwelling animals like the kangaroo rat it is 100 times. This acute hearing warns them of the approach of predators like owls and snakes, even in the dark.

The vibration causes waves in the fluid in the inner ear that pass down the cochlea. These waves stimulate the tiny hair cells to produce nerve impulses that travel via the auditory nerve to the cerebral cortex of the brain where they are interpreted as sound.

To summarise: The route sound waves take as they pass through the ear is: External ear | tympanic membrane | ear ossicles | inner ear |cochlear | hair cells

The hair cells generate a nerve impulse that travels down the auditory nerve to the brain.

Remember that sound waves do not pass along the Eustachian tube. Its function is to equalise the air pressure on either side of the tympanic membrane.

Balance

The vestibular organ of the inner ear helps an animal maintain its posture and keep balanced by monitoring the movement and position of the head. It consists of two structures - the semicircular canals and the otolith organs.

The semicircular canals respond to movement of the body. They tell an animal whether it is moving up or down, left or right. They consist of three canals set in three different planes at right angles to each other so that movement in any direction can be registered. The canals contain fluid and sense cells with fine hairs that project into the fluid. When the head moves the fluid swirls in the canals and stimulates the hair cells. These send nerve impulses along the vestibular nerve to the cerebellum.

Note that the semicircular canals register acceleration and deceleration as well as changes in direction but do not respond to movement that is at a constant speed.

The otolith organs are sometimes known as gravity receptors. They tell you if your head is tilted or if you are standing on your head. They consist of bulges at the base of the semi circular canals that contain hair cells that are covered by a mass of jelly containing tiny pieces of chalk called otoliths. When the head is tilted, or moved suddenly, the otoliths pull on the hair cells, which produce a nerve impulse. This travels down the vestibular nerve to the cerebellum. By coordinating the nerve impulses from the semicircular canals and otolith organs the cerebellum helps the animal keep its balance.

- Receptors for touch, pressure, pain and temperature are found in the skin. Receptors in the muscles, tendons and joints inform the brain of limb position.
- The olfactory organ in the nose responds to chemicals in the air i.e. smell.

- Taste buds on the tongue respond to a limited range of chemicals dissolved in saliva.

- The eyes are the organs of sight. Spherical eyeballs situated in orbits in the skull have walls composed of 3 layers.

- The tough outer sclera protects and holds the shape of the eyeball. At the front it becomes visible as the white of the eye and the transparent cornea that allows light to enter the eye.

- The middle layer is the choroid. It most animals it absorbs stray light rays but in nocturnal animal it is reflective to conserve light. At the front of the eye it becomes the iris with muscles to control the size of the pupil and hence the amount of light entering the eye.

- The inner layer is the retina containing the light receptor cells: the rods for black and white vision in dim light and the cones for colour and detailed vision. Nerve impulses generated by these cells leave the eye for the brain via the optic nerve.

- The lens (with the cornea) helps focus the light rays on the retina. Muscles alter the shape of the lens to allow near and far objects to be focussed.

- Aqueous humour fills the space immediately behind the cornea and keeps it in shape and vitreous humour, a transparent jelly-like substance, fills the space behind the lens allowing light rays to pass through to the retina.

- The ear is the organ of hearing and balance.

- The external pinna helps funnel sound waves into the ear and locate the direction of the sound. The sound waves travel down the external ear canal to the eardrum or tympanic membrane causing it to vibrate. This vibration is transferred to the auditory ossicles of the middle ear which themselves transfer it to the inner ear. Here receptors in the cochlea respond by generating nerve impulses that travel to the brain via the auditory (acoustic) nerve.

- The Eustachian tube connects the middle ear with the pharynx to equalise air pressure on either side of the tympanic membrane.

- The vestibular organ of the inner ear is concerned with maintaining balance and posture. It consists of the semicircular canals and the otolith organs.

Chapter 3
Animal Sexual Behaviour

Animal sexual behaviour takes many different forms, even within the same species. Among animals other than humans, researchers have observed monogamy, promiscuity, sex between species, sexual arousal from objects or places, sex apparently via duress or coercion, copulation with dead animals, homosexual, heterosexual and bisexual sexual behaviour, situational sexual behaviour and a range of other practices. Related studies have noted diversity in sexed bodies and gendered behaviour, such as intersex and transgender animals.

The study of animal sexuality (and primate sexuality especially) is a rapidly developing field. It used to be believed that only humans and a handful of other species performed sexual acts other than for procreation, and that animals' sexuality was instinctive and a simple response to the "right" stimulation (sight, scent). Current understanding is that many species that were formerly believed monogamous have now been proven to be promiscuous or opportunistic in nature; a wide range of species appear both to masturbate and to use objects as tools to help them do so; in many species animals try to give and get sexual stimulation with others where procreation is not the aim; and homosexual behaviour has now been observed among 1,500 species and in 500 of those it is well documented.

Mating System

A mating system is a way in which a group is structured in relation to sexual behaviour. The precise meaning depends upon the context. With respect to higher animals, it specifies which males mate with which females, under which circumstances; recognised animal mating systems include monogamy, polygamy (which includes polygyny, polyandry, and polygynandry) and promiscuity. In plants, it refers to the degree and circumstances of outcrossing. In human sociobiology, the terms have been extended to encompass the formation of relationships such as marriage.

In Animals

The following are some of the mating systems generally recognised in animals:

- *Monogamy:* One male and one female have an exclusive mating relationship. The term "pair bonding" often implies this. It could also mean one or more animals have an exclusive relationship with one or more opposite-sex animals.
- *Polygamy:* Three types are recognised:
  - Polygyny (the most common polygamous mating system in vertebrates so far studied): One male has an exclusive relationship with two or more females
  - Polyandry: One female has an exclusive relationship with two or more males
  - *Polygynandry:* Two or more males have an exclusive relationship with two or more females; the numbers of males and females need not be equal, and in vertebrate species studied so far, the number of males is usually less.
- *Promiscuity:* A member of one sex within the social group mates with any member of the opposite sex.

These mating relationships may or may not be associated with social relationships, in which the sexual partners stay together to become parenting partners. As the alternative term "pair bonding" implies, this is usual in monogamy. In many polyandrous systems, the males and the female stay together to rear the young. In polygynous systems where the number of females paired with each male is low, the male will often stay with one female to help rear the young, while the other females rear their young on their own. In polygynandry, each of the males may assist one female; if all adults help rear all the young, the system is more usually called "communal breeding". In highly polygynous systems, and in promiscuous systems, paternal care of young is rare, or there may be no parental care at all.

It is important to realize that these descriptions are idealised, and that the social partnerships are often easier to observe than the mating relationships. In particular:

- the relationships are rarely exclusive for all individuals in a species. DNA fingerprinting studies have shown that even in pair-bonding, matings outside the pair (extra-pair copulations) occur with fair frequency, and a significant minority of offspring result from them.

- some species show different mating systems in different
  circumstances, for example in different parts of their
  geographical range, or under different conditions of food
  availability
- mixtures of the simple systems described above may occur.

Monogamous Pairing in Animals

Monogamous pairing in animals refers to the natural history of
mating systems in which species pair bond to raise offspring. This is
associated, usually implicitly, with sexual monogamy.

Animals

The evolution of mating systems in animals has received an
enormous amount of attention from biologists. This section briefly
reviews three main findings about the evolution of monogamy in
animals.

The amount of social monogamy in animals varies across taxa,
with over 90% of birds engaging in social monogamy while only 7%
of mammals were known to do the same.

This list is not complete. Other factors may also contribute to the
evolution of social monogamy. Moreover, different sets of factors may
explain the evolution of social monogamy in different species. There
is no one-size-fits-all explanation of why different species evolved
monogamous mating systems.

Sexual Dimorphism

Sexual dimorphism refers to differences in body characteristics
between males and females. A frequently studied type of sexual
dimorphism is body size. Males typically have larger bodies than
females. In some species, however, females have larger bodies than
males. Sexual dimorphism in body size has been linked to mating
behaviour. In polygynous species, males compete for control over
sexual access to females. Large males have an advantage in the
competition for access to females, and they consequently pass their
genes along to a greater number of offspring. This eventually leads
to large differences in body size between males and females. Polygynous
males are often 1.5 to 2.0 times larger in size than females. In
monogamous species, on the other hand, males and females have more
equal access to mates, so there is little or no sexual dimorphism in
body size. From a new biological point of view, monogamy could result

from mate guarding and is engaged as a result of sexual conflict following the hypothesis of mutual destruction.

Some researchers have attempted to infer the evolution of human mating systems from the evolution of sexual dimorphism. Several studies have reported a large amount of sexual dimorphism in *Australopithecus*, an evolutionary ancestor of human beings that lived between 2 and 5 million years ago.

These studies raise the possibility that *Australopithecus* had a polygamous mating system. Sexual dimorphism then began to decrease. Studies suggest sexual dimorphism reached modern human levels around the time of *Homo Erectus* 0.5 to 2 million years ago. This line of reasoning suggests human ancestors started out polygamous and began the transition to monogamy somewhere between 0.5 million and 2 million years ago.

Attempts to infer the evolution of monogamy based on sexual dimorphism remain controversial for three reasons:

- The skeletal remains of *Australopithecus* are quite fragmentary. This makes it difficult to identify the sex of the fossils. Researchers sometimes identify the sex of the fossils by their size, which, of course, can exaggerate findings of sexual dimorphism.

- Recent studies using new methods of measurement suggest *Australopithecus* had the same amount of sexual dimorphism as modern humans. This raises questions about the amount of sexual dimorphism in *Australopithecus*.

- Humans may have been partially unique in that selection pressures for sexual dimorphism might have been related to the new niches that humans were entering at the time, and how that might have interacted with potential early cultures and tool use. If these early humans had a differentiation of gender roles, with men hunting and women gathering, selection pressures in favour of increased size may have been distributed unequally between the sexes.

- Even if future studies clearly establish sexual dimorphism in *Australopithecus*, other studies have shown the relationship between sexual dimorphism and mating system is unreliable. Some polygamous species show little or no sexual dimorphism. Some monogamous species show a large amount of sexual dimorphism.

Studies of sexual dimorphism raise the possibility that early human ancestors were polygamous rather than monogamous. But this line of research remains highly controversial. It may be that early human ancestors showed little sexual dimorphism, and it may be that sexual dimorphism in early human ancestors had no relationship to their mating systems.

Testis Size

The relative sizes of male testes often reflect mating systems. In species with promiscuous mating systems, where many males mate with many females, the testes tend to be relatively large. This appears to be the result of sperm competition. Males with large testes produce more sperm and thereby gain an advantage impregnating females. In polygynous species, where one male controls sexual access to females, the testes tend to be small. One male defends exclusive sexual access to a group of females and thereby eliminates sperm competition.

Studies of primates, including humans, support the relationship between testis size and mating system. Chimpanzees, which have a promiscuous mating system, have large testes compared to other primates. Gorillas, which have a polygynous mating system, have smaller testes than other primates. Humans, which have *a socially monogamous mating system, accompanied by moderate amounts of sexual non-monogamy*, have moderately sized testes. The moderate amounts of sexual non-monogamy in humans may result in a low to moderate amount of sperm competition. Also, notably, in the case of an avowedly sexually monogamous society, the occurrence of sexual nonmonogamy is typically culturally stigmatised, and therefore detecting its prevalence is inherently difficult, if indeed it is at all possible. At best, such statistics can be viewed as general approximations with a wide margin of error.

Although testis size in humans is consistent with the modern pattern of social monogamy accompanied by moderate sexual non-monogamy, this fact reveals little about when the modern pattern evolved. Did Homo Erectus have testes similar in size to modern humans? What about Australopithecus? It is not possible to measure the size of testes in the fossil remains of human ancestors. This limits the usefulness of testis size in understanding the evolution of monogamy in humans.

Monogamy as a Best Response

In species where the young are particularly vulnerable and may benefit from protection by both parents, monogamy may be an optimal strategy. The selection factors in favour of different mating strategies for a species of animal, however, may potentially operate on a large number of factors throughout that animal's life cycle.

For instance, with many species of bear, the female will often drive a male off soon after mating, and will later guard her cubs from him similar to that of other animals after they are born. It is thought that this might be due to fact that too many bears close to one another may deplete an area of food resources for the relatively small but growing cubs.

Monogamy could be social but rarely genetic. Thierry Lodé argued that monogamy should result from conflict of interest between the sexes called sexual conflict. Organised from territory defence and mate guarding, monogamy appears as a response of male for the control of female sexuality, but exclusive monogamy would be rare and the biological evolution would privilege the diversity of sexual behaviours.

Sexual Conflict

Sexual conflict occurs when the two sexes have conflicting optimal fitness strategies concerning reproduction, leading to an evolutionary arms race between males and females. It has primarily been studied in animals, though it can in principle apply to any sexually reproducing organism, such as plants and fungi.

This can be in two forms:

1. Interlocus sexual conflict, where male alleles have conflicting interests with female alleles that are at a different genetic locus. This can be in the form of conflict over parental care: maternal alleles which cause the offspring to require a greater investment from the father, and paternal alleles which cause the offspring to require a greater investment from the mother. Another form is conflict over mating rate. Males frequently have a higher optimal mating rate than females (because in most animal species they invest fewer resources in offspring than females do), and so males have numerous adaptations to induce females to mate with them. Another well-documented example of inter-locus sexual conflict is the seminal fluid of *Drosophila melanogaster*, which up-regulates females' egg-laying rate and reduces her desire to re-mate with another male (serving the male's interests), but also shortens the female's lifespan reducing her fitness.

2. Intralocus sexual conflict, where the same set of alleles in males and females have different optima. i.e. they are expressed differently in the sexes. A classic example is the human hip, where females need larger hips for childbirth. The genes that affect hip size must reach a compromise that is at neither the male optimum nor the female optimum. In some cases, this conflict may be resolved through the differential expression of such loci in males and females, but evidence indicates that intralocus conflict may be an important constraint in the evolution of many traits.

Sexual conflict may lead to antagonistic co-evolution, in which one sex (usually males) evolves a favourable trait which is countered by a countering trait in the other sex. For example, male bean weevils (*Callosobruchus maculatus*) have spiny genitalia that are thought to allow them to copulate for a longer time without getting dislodged and hence transfer more sperm. However, this damages the female and reduces her fitness, so females have evolved the counter adaptation of kicking at males during mating, which reduces the time spent in copulation.

Some regard sexual conflict as a subset of sexual selection (which was traditionally regarded as mutualistic), while others suggest it is a separate evolutionary phenomenon.

Manifestations

There are a wide variety of manifestations of sexual conflict, occurring in a broad range of taxa, and showing extreme examples such as the well-known phenomenon of sexual cannibalism among insects, usually demonstrated by the female. One way of sorting these manifestations is by temporal relation to a given reference point, e.g. mating, fertilization, or birth.

Infanticide (Zoology)

In animals, infanticide involves the killing of young offspring by a mature animal of its own species, and is studied in zoology, specifically in the field of ethology. Ovicide is the analogous destruction of eggs. Although human infanticide has been widely studied, the practice has been observed in many other species throughout the animal kingdom. These include microscopic rotifers, insects, fish, amphibians, birds and mammals. Infanticide can be practiced by both males and females.

Infanticide caused by sexual conflict has the general theme of the killer (often male) becoming the new sexual partner of the victim's parent, which would otherwise be unavailable. This represents a gain in fitness by the killer, and a loss in fitness by the parents of the offspring killed. This is a type of evolutionary struggle between the two sexes, in which the victim sex may have counter-adaptations that reduce the success of this practice. It may also occur for other reasons, such as the struggle for food between females. In this case individuals may even kill closely related offspring.

Filial infanticide occurs when a parent kills its own offspring. This sometimes involves consumption of the young themselves, which is termed filial cannibalism. The behaviour is widespread in fishes, and is seen in terrestrial animals as well. Human infanticide has been recorded in almost every culture. A unique aspect of human infanticide is selective killing based on gender.

Background

Infanticide only came to be seen as a significant occurrence in nature quite recently. At the time it was first seriously treated by Yukimaru Sugiyama, infanticide was attributed to stress causing factors like overcrowding and captivity, and was considered pathological and maladaptive. Classical ethology held that conspecifics (members of the same species) rarely killed each other. By the 1980s it had gained much greater acceptance. Possible reasons it was not treated as a serious natural phenomenon include its abhorrence to people, the popular group and species selectionist notions of the time (the idea that individuals behave for the good of the group or species; compare with gene-centred view of evolution), and the fact that it is very difficult to observe in the field.

Infanticide Involving Sexual Conflict

This form of infanticide represents a struggle between the sexes, where one sex exploits the other, much to the latter's disadvantage. It is usually the male who benefits from this behaviour, though in cases where males play similar roles to females in parental care the victim and perpetrator may be reversed.

By Males

Hanuman langurs (or gray langurs) are an Old World monkey found in India. They are a social animal, living in groups. Each group is generally dominated by a single male, with many females, though the male must struggle for control of the group with other males. When a male tries to take over a group, there is a violent struggle with the existing male. If successful in overthrowing the previous male, infants of the females are then killed. This infanticidal period is limited to the window just after the group is taken over. Cannibalism, however, has not been observed in this species.

This behaviour not only reduces intraspecific competition between the incumbent's offspring and those of other males and increases the parental investment afforded to their own young, but also allows females to become sexually receptive sooner. This is because females of this species, as well as many other mammals, do not ovulate during the period in which they produce milk. It then becomes easier to see how this behaviour could have evolved. If a male kills a female's young, they stop lactating and are able to become pregnant again. As males are in a constant struggle to protect their group, those that express infanticidal behaviour will contribute a larger portion to future gene pools.

Similar behaviour is also seen in male lions, among other species, who also kill young cubs, allowing them to impregnate the females. Unlike langurs, male lions live in small groups, which cooperate to take control of a pride from an existing group. They will attempt to kill any cubs that are roughly 9 months old or less, though as in other species, the female will attempt to defend their cubs viciously. Males have, on average, only a two year window in which to pass on their genes, and female lions only give birth once every two years, so the selective pressure from them to behave like this is strong. In fact it is estimated that a quarter of cubs dying in the first year of their life are victims of infanticide.

Male mice show great variation in behaviour over time. After fertilizing a female, they become aggressive towards mouse pups for three weeks, killing any they come across. After this period however, their behaviour changes dramatically, and they become paternal, caring for their own offspring. This lasts for almost two months, but afterwards they become infanticidal once more.

It is no coincidence here that the female gestation period is three weeks as well, or that it takes roughly two months for pups to become fully weaned and leave their nest. The proximate mechanism that allows for the correct timing of these periods involves circadian rhythms, each day and night cycle affecting the mouse's internal neural physiology, and disturbances in the duration of these cycles results in different periods of time between behaviours. The adaptive value of this behaviour switching is twofold; infanticide removes competitors for when the mouse does have offspring, and allows the female victims to be impregnated earlier than if they continued to care for their young, as mentioned above.

Gerbils, on the other hand, no longer commit infanticide once they have paired with a female, but actively kill and eat other offspring when young. The females of this species behave much like male mice, hunting down other litters except when rearing their own.

Prospective Infanticide

Prospective infanticide is a subset of sexual competition infanticide in which young born *after* the arrival of the new male are killed. This is less common than infanticide of existing young, but can still increase fitness in cases where the offspring could not possibly have been fathered by the new mate, i.e. one gestation or fertility period. This is known to occur in lions and langurs, and has also been observed in other species such as house wrens. In birds, however, the situation is more complex, as female eggs are fertilized one at a time, with a 24-hour delay between each. Males may destroy clutches laid 12 days or more after their arrival, though their investment of around 60 days of parental care is large, so a high level of parental certainty is needed.

By Females

Females are also known to display infanticidal behaviour. This may appear unexpected, as the conditions described above do not apply. Males are not always an unlimited resource though - in some species, males provide parental care to their offspring, and females may compete indirectly with others by killing their offspring, freeing up the limiting resource that the males represent. This has been documented in research by Stephen Emlen and Natalie Demong on wattled jacanas (*Jacana jacana*), a tropical wading bird. With the wattled jacana, it is exclusively the male sex that broods, while females defend their territory. In this experiment Demong and Emlen found that removing females from a territory resulted in nearby females attacking the chicks of the male in most cases, evicting them from their nest. The males then fertilized the offending females and cared for their young. Emlen describes how he "shot a female one night, and [...] by first light a new female was already on the turf. I saw terrible things—pecking and picking up and throwing down chicks until they were dead. Within hours she was soliciting the male, and he was mounting her the same day. The next night I shot the other female, then came out the next morning and saw the whole thing again."

Infanticide is also seen in giant water bugs. *Lethocerus deyrollei* is a large and nocturnal predatory insect found in still waters near vegetation. In this species the males take care of masses of eggs by keeping them hydrated with water from their bodies. Without a male caring for the eggs like this, they become desiccated and will not hatch. In this species, males are a scarce resource that females must sometimes compete for. Those that cannot find a free male often stab the eggs of a brooding one. As in the above case, males then fertilize this female and care for her eggs.

Noritaka Ichikawa has found that males only moisten their eggs during the first 90 seconds or so, after which all of the moisture on their bodies has evaporated. However, they guard the egg masses for as long as several hours at a time, when they could be hunting prey. They do not seem to prevent further evaporation by staying guard, as males that only guarded the nest for short periods were seen to have similar hatching rates in a controlled experiment where there were no females present. It seems rather that males are more successful in avoiding infanticidal females when they are out of the water with their eggs, which might well explain the ultimate cause of this behaviour.

Resource Competition

Black-tailed prairie dogs are colonially-living, harem-polygynous squirrels found mainly in the United States. Their living arrangement involves one male living with four or so females in a territory defended by all individuals, and underground nesting. Black-tails only have one litter per year, and are in estrous for only a single day around the beginning of spring.

A seven-year natural experiment by John Hoogland and others from the University of Princeton revealed that infanticide is widespread in this species, including infanticide from invading males and immigrant females, as well as occasional cannibalism of an individual's own offspring.

The surprising finding of the study was that by far the most common type of infanticide involved the killing of close kin's offspring. This seems illogical, as kin selection favours behaviours that promote the well-being of closely related individuals. It was postulated that this form of infanticide is more successful than trying to kill young in nearby groups, as the whole group must be bypassed in this case, while within a group only the mother need be evaded. Marauding behaviour is evidently adaptive, as infanticidal females had more and healthier young than others, and were heavier themselves as well. This behaviour appears to reduce competition with other females for food, and future competition among offspring.

Similar behaviour has been reported in the meerkat (*Suricata suricatta*), including cases of females killing their mother's, sister's, and daughter's offspring. Infanticidal raids from neighbouring groups also occurred.

Other

Bottlenose dolphins have been reported to kill their young through impact injuries. Dominant male langurs tend to kill the existing young upon taking control of a harem. There has been sightings of infanticide in the leopard population.

Costs and Defences

Costs of the Behaviour: While it may be beneficial for some species to behave this way, infanticide is not without risks to the perpetrator. Having already exhausted energy and perhaps sustained serious wounds in a fight with another male, attacks from females who vigorously defend their offspring may be telling for harem-polygynous males, with a risk of infection. It is also energetically costly to pursue a mother's young, which may try to escape.

Costs of the behaviour described in prairie dogs include the risk to an individual of losing their own young while killing another's, not to mention the fact that they are killing their own relatives. In a species where infanticide is common, perpetrators may well be victims themselves in the future, such that they come out no better off; but as long as an infanticidal individual gains in reproductive output by its behaviour, it will tend to become common. Further costs of the behaviour in general may be induced by counter-strategies evolved in the other sex, as described below.

As a Cost of Social Behaviour

Taking a broader view of the black-tailed prairie dog situation, infanticide can be seen as a cost of social living. If each female were to have her own private nest away from others, she would be much less likely to have her infants killed when absent. This, and other costs such as increased spread of parasites, must be made up for by other benefits, such as group territory defence and increased awareness of predators.

An avian example published in *Nature* is acorn woodpeckers. Females nest together, possibly because those nesting alone have their eggs constantly destroyed by rivals. Even so, eggs are consistently removed at first by nest partners themselves, until the entire group lays on the same day. They then cooperate and incubate the eggs as a group, but by this time a significant proportion of their eggs have been lost because of this ovicidal behaviour.

Counter-strategies

Because this form of infanticide reduces the fitness of killed individuals' parents, animals have evolved a range of counter-strategies against this behaviour. These may be divided into two very different classes - those that tend to prevent infanticide, and those that minimise losses.

Loss Minimisation

Some females abort or resorb their own young while they are still in development after a new male takes over; this is known as the Bruce effect. This may prevent their young from being killed after birth, saving the mother wasted time and energy. However, this strategy also benefits the new male. In mice this can occur by the proximate mechanism of the female smelling the odor of the new male's urine.

Preventative Adaptations

Infanticide in burying beetles may have led to male parental care. In this species males often cooperate with the female in preparing a piece of carrion, which is buried with the eggs and eaten by the larvae when they hatch. Males may also guard the site alongside the female. It is apparent from experiments that this behaviour does not provide their young with any better nourishment, nor is it of any use in defending against predators. However, other burying bugs may try to take their nesting space. When this occurs, a male-female pair is over twice as successful in nest defence, preventing the ovicide of their offspring.

Female langurs may leave the group with their young alongside the outgoing male, and others may develop a *false estrous* and allow the male to copulate, deceiving him into thinking she is actually sexually receptive. Females may also have sexual liaisons with other males. This promiscuous behaviour is adaptive, because males will not know whether it is their own offspring they are killing or not, and may be more reluctant or invest less effort in infanticide attempts.

Lionesses cooperatively guard against scouting males, and a pair were seen to violently attack a male after it killed one of their young. Resistance to infanticide is also costly, though: for instance, a female may sustain serious injuries in defending her young. At times it is simply more advantageous to submit than to fight.

Infanticide by Parents and Caregivers

Filial infanticide occurs when a parent kills its own offspring. Both male and female parents have been observed to do this, as well as sterile worker castes in some eusocial animals.

Maternal

Maternal infanticide occurs when newborn offspring are killed by their mother. This is sometimes seen in pigs, a behaviour known as savaging, which affects up to 5% of gilts. Similar behaviour has been observed in various animals such as rabbits and burying beetles.

Paternal

Paternal infanticide—where fathers eat their own offspring—may also occur. When young bass hatch from the spawn, the father guards the area, circling around them and keeping them together, as well as providing protection from would-be predators. After a few days, most of the fish will swim away. At this point the male's behaviour changes: instead of defending the stragglers, he treats them as any other small prey, and eats them.

Worker Caste Killing Young

Honey bees may become infected with a bacterial disease called foul brood, which attacks the developing bee larva while still living in the cell. Some hives however have evolved a behavioural adaptation that resists this disease: the worker bees selectively kill the infected individuals by removing them from their cells and tossing them out of the hive, preventing it from spreading.

The genetics of this behaviour are quite complex. Experiments by Rothenbuhler showed that the 'hygienic' behaviour of the queen was lost by crossing with a non-hygienic drone. This means that the trait must be recessive, only being expressed when both alleles contain the gene for hygienic behaviour. Furthermore, the behaviour is dependent on two separate loci. A backcross produced a mixed result. The hives of some offspring were hygienic, while others were not. There was also a third type of hive where workers removed the wax cap of the infected cells, but did nothing more. What was not apparent was the presence of a fourth group who threw diseased larvae out of the hive, but did not have the uncapping gene. This was suspected by Rothenbuhler however, who manually removed the caps, and found some hives proceeded to clear out infected cells.

Humans and Infanticide

Perceptions and Importance: Like many of the darker aspects of nature, infanticide is a subject that many find discomforting. Cornell University ethologist Glenn Hausfater states that "infanticide has not received much study because it's a repulsive subject [...] Many people regard it as reprehensible to even think about it."

Research into infanticide in animals is in part motivated by the desire to understand human behaviours, such as child abuse. Hausfater explains that researchers are "trying to see if there's any connection between animal infanticide and child abuse, neglect and killing by humans. We just don't know yet what the connections are."

Infanticide by Humans

Infanticide has been, and still is, practiced by some cultures, groups, or individuals. It is often, but not always, the mother who commits the act. In many past societies, certain forms of infanticide were considered permissible, whereas in most modern societies the practice is considered immoral and criminal. Nonetheless, it still takes place — in the Western world usually because of the parent's mental illness or violent behaviour, and in some poor countries as a form of population control, sometimes with tacit societal acceptance. Female infanticide, a form of sex-selective infanticide, is more common than the killing of male offspring.

During pregnancy, there is a two-way traffic of immunologically active cell lines through the placenta. Fetal lymphocyte lines may survive in women even decades after giving birth. These cells may serve the adaptive interest of the mother, however, they may also serve conflicting interests of the fetus or those of the father. This mixture of shared or contradicting interests have been hypothesized to give rise to diseases like autoimmune diseases, infertility, and habitual spontaneous abortion (i.e., early miscarriage) in humans.

Harem (Zoology)

The term harem is used in zoology to describe the social organisation of certain herbivore species, such as those in the Hominidae and Equidae families, into groups of females and young surrounding a single dominant male. Non dominant males may form bachelor groups, or may occasionally be permitted to remain on the periphery of the harem.

Social Monogamy

Social monogamy refers to a male and female's social living arrangement (e.g., shared use of a territory, behaviour indicative of a social pair, and/or proximity between a male and female) without inferring any sexual interactions or reproductive patterns. In humans, social monogamy equals monogamous marriage. Sexual monogamy is defined as an exclusive sexual relationship between a female and a male based on observations of sexual interactions. Finally, the term genetic monogamy is used when DNA analyses can confirm that a female-male pair reproduce exclusively with each other. A combination of terms indicates examples where levels of relationships coincide, e.g., sociosexual and sociogenetic monogamy describe corresponding social and sexual, and social and genetic monogamous relationships, respectively.   —Reichard, 2003, p. 4

Whatever makes a pair of animals socially monogamous does not necessarily make them sexually or genetically monogamous. Social monogamy, sexual monogamy, and genetic monogamy can occur in different combinations.

Social monogamy is relatively rare in the animal kingdom. The actual incidence of social monogamy varies greatly across different branches of the evolutionary tree. Over 90 percent of avian species are socially monogamous.

This stands in contrast to mammals. Only 3 percent of mammalian species are socially monogamous, although up to 15 percent of primate species are socially monogamous. Social monogamy has also been observed in reptiles, fish, and insects.

Sexual monogamy is also rare among animals. Many socially monogamous species engage in extra-pair copulations, making them sexually non-monogamous. For example, while over 90% of birds are socially monogamous, "on average, 30 percent or more of the baby birds in any nest [are] sired by someone other than the resident male." Patricia Adair Gowaty has estimated that, out of 180 different species of socially monogamous songbirds, only 10% are sexually monogamous.

The incidence of genetic monogamy, determined by DNA fingerprinting, varies widely across species. For a few rare species, the incidence of genetic monogamy is 100 percent, with all offspring genetically related to the socially monogamous pair. But genetic monogamy is strikingly low in other species. Barash and Lipton note:

> The highest known frequency of extra-pair copulations are found among the fairy-wrens, lovely tropical creatures technically known as *Malurus splendens* and *Malurus cyaneus*. More than 65 percent of all fairy-wren chicks are fathered by males outside the supposed breeding group.   —Barash & Lipton, 2001, p. 12

Such low levels of genetic monogamy have surprised biologists and zoologists, forcing them to rethink the role of social monogamy in evolution. They can no longer assume social monogamy determines how genes are distributed in a species. The lower the rates of genetic monogamy among socially monogamous pairs, the less of a role social monogamy plays in determining how genes are distributed among offspring.

Polygamy

Polygamy is defined as a mating structure in which a single individual of one gender has exclusive access to several individuals of the opposite gender. It takes two main forms – polygyny and polyandry. As polygyny is the most common form of polygamy among vertebrates (including humans, to some extent), it has been studied far more extensively than polyandry.

Polygyny

In some species, notably those with harem-like structures, only one of a few males in a group of females will mate. Technically, polygyny in sociobiology and zoology is defined as a system in which a male has a relationships with more than one female, but the females are predominantly bonded to a single male. Should the active male be driven out, killed, or otherwise removed from the group, in a number of species the new male will ensure that breeding resources are not wasted on another male's young. The new male may achieve this in many different ways, including:

- *Competitive Infanticide:* In lions, hippopotamuses, and some monkeys, the new male will kill the offspring of the previous alpha male to cause their mothers to become receptive to his sexual advances since they are no longer nursing.

- *Harassment to Miscarriage:* Amongst wild horses and baboons, the male will "systematically harass" pregnant females until they miscarry.

- *Pheromone based Spontaneous Abortion:* In some rodents such as mice, a new male with a different scent will cause females who are pregnant to spontaneously fail to implant recently fertilized eggs. This does not require contact; it is mediated by scent alone. It is known as the Bruce-Parkes effect.

Promiscuity

Two examples of systems in primates are promiscuous mating chimpanzees and bonobos. These species live in social groups consisting of several males and several females. Each female copulates with many males, and vice versa. In bonobos, the amount of promiscuity is particularly striking because bonobos use sex to alleviate social conflict as well as to reproduce.

Seasonal Breeder

Seasonal breeders are animal species that successfully mate only during certain times of the year. These times of year allow for the births at a time optimal for the survival of the young in terms of factors such as temperature, food and water. Related sexual interest and behaviours are expressed and accepted only during this period.

Female seasonal breeders will have one or more estrus cycles only when she is "in season" or fertile and receptive to mating. At other times of the year, they will be anestrus.

Similarly, male seasonal breeders may exhibit changes in testosterone levels, testes weight and fertility depending on the time of year.

Seasonal breeders are distinct from opportunistic breeders, which mate whenever the conditions of their environment become favourable, and continuous breeders like humans that mate year-round.

Physiology

The hypothalamus is considered to be the central control for reproduction. Hence, factors that determine when a seasonal breeder will be ready for mating affect this tissue.

This is achieved specifically through changes in the production of the hormone GnRH. GnRH in turn transits to the pituitary where it promotes the secretion into the bloodstream of the gonadotropin LH, a pituitary hormone critical for reproductive function and behaviour. Changes in gonadotropin secretion initiate the end of anestrus in females.

Factors that Determine Time of Fertility

Photoperiod: When a seasonal breeder is ready for mating is strongly regulated by length of day (photoperiod) and thus season. Photoperiod likely affects the seasonal breeder through changes in melatonin secretion by the pineal gland that ultimately alter GnRH release by the hypothalamus.

Hence, seasonal breeders can be divided into groups based on when they are fertile. "Long day" breeders cycle when days get longer (spring) and are anestrus in fall and winter. "Short day" breeders cycle when the length of daylight shortens (fall) and are anestrus in spring and summer. Domestication has allowed cattle and swine to be liberated from breeding seasonality. Day length variations with latitude can also impact breeding. For instance, sheep and goats in tropical climes may breed throughout the year while those in more polar arctic areas may have a shortened season.

Females are generally more sensitive to changes in day length. For instance, unlike mares, stallions remain fertile year-round, suffering only some declines in sexual behaviour and sperm production out of season.

Other Factors

Other factors that affect breeding time include the presence of a ready and available mate. For instance, the presence of a fertile male will induce an estrus cycle in a doe shortly after introduction.

Further environmental factors can include nutrition, chemosensory and hormonal cues. Weight and age are other factors.

Partial List of Seasonal Breeders

Many non-mammals are seasonal breeders, such as many birds and fish. Here is partially listed those that are mammals.

Long Day Breeders

- Ring-tailed lemur
- Horse
- Hamster
- Groundhog
- Mink.

Short Day Breeders

- Sheep
- Goat

- Fox
- Deer, Red Deer
- Elk
- Moose.

Summer Breeders

- Ruffed lemur (May - July)
- Select species of hamster, vole and mouse.

Interpretation Bias

The field of study of sexuality in non-human species has been a long standing taboo, with researchers either failing to observe or miscategorising and misdescribing sexual behaviour which does not meet their preconceptions. More current research provides views such as that of the Natural History Museum at the University of Oslo, which in 2006 held an exhibition on animal sexuality:

Many researchers have described homosexuality as something altogether different from sex. They must realise that animals can have sex with who they will, when they will and without consideration to a researcher's ethical principles.

An example of overlooking behaviour relates to descriptions of giraffe mating:

When nine out of ten pairings occur between males, "[e]very male that sniffed a female was reported as sex, while anal intercourse with orgasm between males was only [categorised as] 'revolving around' dominance, competition or greetings.

Sex for Pleasure

It is a common myth that animals do not (as a rule) have sex for pleasure, or alternatively that humans, pigs (and perhaps dolphins and one or two species of primate) are the only species which do. This is sometimes formulated "animals mate only for reproduction".

Science cannot conclusively say at present what animals do or do not find "pleasurable", a question considered in more depth under Emotion in animals. The urban considers this particular view in depth. Its conclusions are broadly that the statement is true, but only using a *very specific definition* of "sex for pleasure", in which sexual acts tied to a reproductive cycle or for which an alternative explanation can be asserted, are ignored, as is all sexual activity that does not involve penetration. Animals put themselves at risk to engage in sex, and as a result, most species have evolved sexual signals (usually scent and behaviour) to indicate the presence of receptive periods. During these, sex is sought, and outside these it is usually not sought (or is sought but not permitted). Snopes comments that this is not in fact a reflection of whether sex is pleasurable or not, but rather a reflection of whether individuals have sex at arbitrary times. They conclude:

> "Of course, we have to make many seemingly artificial distinctions to arrive at our conclusion. Animals other than humans have no awareness that their sexual activities are connected with reproduction: They engage in sex because they're biologically driven to do so, and if the fulfillment of their urges produces a physical sensation we might appropriately call 'pleasure,' it isn't the least bit affected by the possibility (or impossibility) of producing offspring. We are also discounting cases in which animals do engage in sex even though reproduction is an impossibility because we claim there are other 'purposes' (of which the animals themselves are unaware) at play. (For example, the females of some species of birds will invite males to mate with them even after they have laid their eggs, but we ascribe a purpose to this behaviour: this is a biological "trick" to fool males into caring for hatchlings they didn't father.) We also employ subjective terms such as 'willingly' and 'regularly' in claiming that bonobos and dolphins are the only other animals who "willingly (and regularly) engage in sex with each other" ... and even then it may be the case that these species have some other 'purpose' for doing so that we haven't yet discovered..."

A 2006 Danish Animal Ethics Council report which examined current knowledge of animal sexuality in the context of legal queries concerning sexual acts by humans, has the following comments, primarily related to domestically common animals:

Even though the evolution-related purpose of mating can be said to be reproduction, it is not actually the creating of offspring which originally causes them to mate. It is probable that they mate because they are motivated for the actual copulation, and because this is connected with a positive experience. It is therefore reasonable to assume that there is some form of pleasure or satisfaction connected with the act. This assumption is confirmed by the behaviour of males, who in the case of many species are prepared to work to get access to female animals, especially if the female animal is in oestrus, and males who for breeding purposes are used to having sperm collected become very eager, when the equipment they associate with the collection is taken out.

There is nothing in female mammals' anatomy or physiology, that contradicts that stimulation of the sexual organs and mating is able to be a positive experience. For instance, the clitoris acts in the same way as with women, and scientific studies have shown that the success of reproduction is improved by stimulation of clitoris on (among other species) cows and mares in connection with insemination, because it improves the transportation of the sperm due to contractions of the inner genitalia. This probably also concerns female animals of other animal species, and contractions in the inner genitals are seen e.g. also during orgasm for women. It is therefore reasonable to assume that sexual intercourse may be linked with a positive experience for female animals.

Types of Activity

Autoeroticism or Masturbation: It appears that many animals, both male and female, masturbate, both when partners are available and otherwise.

For example, Petplace comments in its guide on assessing potential breeding stock purchases: *"Masturbation is a normal behaviour in all stallions that does not reduce semen production or performance in the breeding shed"* Likewise a review from the University of Pennsylvania School of Veterinary Medicine says:

> the behaviour known within the horse breeding industry as masturbation. *This involves normal periodic erections and penile movements. This behaviour, both from the descriptive field studies cited above and in extensive study of domestic horses, is now understood as normal, frequent behaviour of male equids. Attempting to inhibit or punish masturbation, for example by tying a brush to the area of the flank underside where the penis rubs into contact with the underside, which is still a common practice of horse managers regionally around the world, often leads to increased masturbation and disturbances of normal breeding behaviour.*
>
> —Sue M. McDonnell, *Sexual Behaviour – Current Topics in Applied Ethology and Clinical Methods*

Castration does not prevent masturbation, as it is observed in geldings. Masturbation is common in both mares and stallions, before and after puberty. Sexologist Havelock Ellis in his 1927 "Studies in the Psychology of Sex" identified bulls, goats, sheep, camels and elephants as species known to practice autoeroticism, adding of some other species:

> *I am informed by a gentleman who is a recognised authority on goats, that they sometimes take the penis into the mouth and produce actual orgasm, thus practicing auto-fellatio. As regards ferrets ... "if the bitch, when in heat, cannot obtain a dog [ie, male ferret] she pines and becomes ill. If a smooth pebble is introduced into the hutch, she will masturbate upon it, thus preserving her normal health for one season. But if this artificial substitute is given to her a second season, she will not, as formerly, be content with it." Blumenbach observed a bear act somewhat similarly on seeing other bears coupling, and hyenas, according to Ploss and Bartels, have been seen practicing mutual masturbation by licking each other's genitals.*

In his 1999 book, *Biological exuberance*, Bruce Bagemihl PhD documents (p. 71, 209–210) that:

> *Autoeroticism also occurs widely among animals, both male and female. A variety of creative techniques are used, including genital stimulation using the hand or front paw (primates, Lions), foot (Vampire Bats, primates), flipper (Walruses), or tail (Savanna Baboons), sometimes accompanied by stimulation of the nipples (Rhesus Macaques, Bonobos); auto-fellating or licking, sucking and/or nuzzling by a male of his own penis (Common Chimpanzees, Savanna Bonobos, Vervet Monkeys, Squirrel Monkeys, Thinhorn Sheep, Bharal, Aovdad, Dwarf Cavies); stimulation of the penis by flipping or rubbing it against the belly or in its own sheath (White-tailed and Mule Deer, Zebras and Takhi); spontaneous ejaculations (Mountain Sheep, Warthogs, Spotted Hyenas); and stimulation of the genitals using inanimate objects (found in several primates and cetaceans).*

Many birds masturbate by mounting and copulating with tufts of grass, leaves or mounds of earth, and some mammals such as primates and dolphins also rub their genitals against the ground or other surfaces to stimulate themselves.

Autoeroticism in female mammals, as well as heterosexual and homosexual intercourse (especially in primates), often involves direct or indirect stimulation of the clitoris. This organ is present in the females of all mammalian species and several other animal groups.

And that:

> *Apes and Monkeys use a variety of objects to masturbate with and even deliberately create implements for sexual stimulation [...] often in highly creative ways.*

Petter Bøckman of the Natural History Museum at the University of Oslo commented (in respect of a 2006 exhibition on homosexuality in the animal kingdom) that:

> *Masturbation is common in the animal kingdom ... We have a Darwinist mentality that all animals only have sex to procreate. But there are plenty of animals who will masturbate when they have nothing better to do. Masturbation has been observed among primates, deer, killer whales and penguins, and we're talking about both males and females. They rub themselves against stones and roots. Orangutans are especially inventive. They make dildos of wood and bark.*

Oral Sex

Animals of several species are documented as engaging in both autofellatio and oral sex. Although easily confused by lay-people, this is a separate and sexually oriented behaviour, distinct from non-sexual grooming or the investigation of scents.

Auto-fellatio or oral sex in animals is documented in goats, primates, hyaenas, bats and sheep.

Contraceptive Sex

Among monkeys, Lionel Tiger and Robin Fox conducted a study on how Depo-Provera contraceptives lead to decreased male attractiveness to females and eventually to male homosexuality. Janet E. Smith summarizes the findings as follows:

The study in the early 70s involved a tribe of monkeys. The alpha monkey of this tribe, named Austin, chose three female monkeys to be his exclusive sexual partners. Austin had a grand time with these three female monkeys. Then the researchers injected Austin's three females with the contraceptive Depo-Provera. Austin stopped having sex with them and chose other female monkeys to be his sexual partners. Then they contracepted all of the females in the tribe. The males stopped having sex with the females and started behaving in a turbulent and confused manner.

Homosexual Behaviour in Animals

Homosexual behaviour in animals refers to the documented evidence of homosexual and bisexual behaviour in non-human species. Such behaviours include sex, courtship, affection, pair bonding, and parenting among same sex animals. A 1999 review by researcher Bruce Bagemihl shows that homosexual behaviour has been observed in close to 1,500 species, ranging from primates to gut worms, and is well documented for 500 of them.

Animal sexual behaviour takes many different forms, even within the same species. The motivations for and implications of these behaviours have yet to be fully understood, since most species have yet to be fully studied. According to Bagemihl, "the animal kingdom does it with much greater sexual diversity — including homosexual, bisexual and nonreproductive sex — than the scientific community and society at large have previously been willing to accept." Current research indicates that various forms of same-sex sexual behaviour are found throughout the animal kingdom. A new review made in 2009 of existing research showed that same-sex behaviour is a nearly universal phenomenon in the animal kingdom, common across species.

Homosexual behaviour is best known from social species. According to geneticist Simon Levay in 1996, "Although homosexual behaviour is very common in the animal world, it seems to be very uncommon that individual animals have a long-lasting predisposition to engage in such behaviour to the exclusion of heterosexual activities. Thus, a homosexual orientation, if one can speak of such thing in animals, seems to be a rarity. One species in which exclusive homosexual orientation occurs, however, is that of domesticated sheep (*Ovis aries*). "About 10% of rams (males) refuse to mate with ewes (females) but do readily mate with other rams."

The observation of homosexual behaviour in animals can be seen as both an argument for and against the acceptance of homosexuality in humans, and has been used especially against the claim that it is a *peccatum contra naturam* ('sin against nature'). For instance, homosexuality in animals was cited in the United States Supreme

Court's decision in *Lawrence v. Texas* which struck down the sodomy laws of 14 states. Whether animal sexuality has logical, ethical, or moral implications in human sexuality is also a source of debate.

Applying the Term "Homosexual" to Animals

The term *homosexual* was coined by Karl-Maria Kertbeny in 1868 to describe same-sex sexual attraction and sexual behaviour in humans. Its use in animal studies has been controversial for two main reasons: animal sexuality and motivating factors have been and remain poorly understood, and the term has strong cultural implications in western society that are irrelevant for species other than humans. Thus homosexual behaviour has been given a number of terms over the years. When describing animals, the term "homosexual" is preferred over "gay", "lesbian" and other terms currently in use, as these are seen as even more bound to human homosexuality.

Animal preference and motivation is always inferred from behaviour. In wild animals, researchers will as a rule not be able to map the entire life of an individual, and must infer from frequency of single observations of behaviour. The correct usage of the term *homosexual* is that an animal *exhibits homosexual behaviour* or even *same-sex sexual behaviour*; however, this article conforms to the usage by modern research applying the term *homosexuality* to all sexual behaviour (copulation, genital stimulation, mating games and sexual display behaviour) between animals of the same sex. In most instances, it is presumed that the homosexual behaviour is but part of the animal's overall sexual behavioural repertoire, making the animal "bisexual" rather than "homosexual" as the terms are commonly understood in humans, but cases of homosexual preference and exclusive homosexual pairs are known.

Research on Homosexual Behaviour in Animals

The presence of same-sex sexual behaviour was not 'officially' observed on a large scale until recent times, possibly due to observer bias caused by social attitudes to same-sex sexual behaviour, innocent confusion, or even from a fear of "being ridiculed by their colleagues." Georgetown University biologist Janet Mann states "Scientists who study the topic are often accused of trying to forward an agenda, and their work can come under greater scrutiny than that of their colleagues who study other topics." They also noted "Not every sexual act has a reproductive function ... that's true of humans and non-humans." It appears to be widespread amongst social birds and mammals, particularly the sea mammals and the primates. The true extent of homosexuality in animals is not known.

While studies have demonstrated homosexual behaviour in a number of species, Petter Bøckman, the scientific advisor of the exhibition Against Nature? In 2007, speculated that the true extent of the phenomenon may be much larger than was then recognised:

No species has been found in which homosexual behaviour has *not* been shown to exist, with the exception of species that never have sex at all, such as sea urchins and aphis. Moreover, a part of the animal kingdom is hermaphroditic, truly bisexual. For them, homosexuality is not an issue.

An example of overlooking homosexual behaviour is noted by Bruce Bagemihl describing mating giraffes where nine out of ten pairings occur between males.

Every male that sniffed a female was reported as sex, while anal intercourse with orgasm between males was only "revolving around" dominance, competition or greetings.

Some researchers believe this behaviour to have its origin in male social organisation and social dominance, similar to the dominance traits shown in prison sexuality. Others, particularly Joan Roughgarden, Bruce Bagemihl, Thierry Lodé and Paul Vasey suggest the social function of sex (both homosexual and heterosexual) is not necessarily connected to dominance, but serves to strengthen alliances and social ties within a flock. Others have argued that social organisation theory is inadequate because it cannot account for some homosexual behaviours, for example, penguin species where same-sex individuals mate for life and refuse to pair with females when given the chance. While reports on many such mating scenarios are still only anecdotal, a growing body of scientific work confirms that permanent homosexuality occurs not only in species with permanent pair bonds, but also in non-monogamous species like sheep.

One report on sheep cited below states:

Approximately 8% of rams exhibit sexual preferences, that is even when given a choice for male partners (male-oriented rams) in contrast to most rams, which prefer female partners (female-oriented rams). We identified a cell group within the medial preoptic area/ anterior hypothalamus of age-matched adult sheep that was significantly larger in adult rams than in ewes.

In fact, apparent homosexual individuals are known from all of the traditional domestic species, from sheep, cattle and horses to cats, dogs and budgerigars.

Genetic and Physiological Basis for Homosexual Animal Behaviour

Researchers found that disabling the (fucose mutarotase) FucM gene in laboratory mice – which influences the levels of estrogen to which the brain is exposed – caused the female mice to behave as if they were male as they grew up. "The mutant female mouse underwent a slightly altered developmental programme in the brain to resemble the male brain in terms of sexual preference" said Professor Chankyu Park of the Korea Advanced Institute of Science and Technology in Daejon, South Korea, who led the research. His most recent findings have been published in the BMC Genetics journal on July 7, 2010.

In March 2011, research shows that serotonin is involved in the mechanism of sexual orientation of mice. Some selected species and groups

Birds

Black Swans: An estimated one-quarter of all black swans pairings are homosexual and they steal nests, or form temporary threesomes with females to obtain eggs, driving away the female after she lays the eggs. More of their cygnets survive to adulthood than those of different-sex pairs, possibly due to their superior ability to defend large portions of land. The same reasoning has been applied to male flamingo pairs raising chicks.

Gulls: Studies have shown that 10 to 15 percent of female western gulls in some populations in the wild exhibit homosexual behaviour.

Ibises: Research has shown that the environmental pollutant methylmercury can increase the prevalence of homosexual behaviour in male American White Ibis. The study involved exposing chicks in varying dosages to the chemical and measuring the degree of homosexual behaviour in adulthood. The results discovered was that as the dosage was increased the likelihood of homosexual behaviour also increased. The endocrine blocking feature of mercury has been suggested as a possible cause of sexual disruption in other bird species.

Mallards: Mallards form male-female pairs only until the female lays eggs, at which time the male leaves the female. Mallards have rates of male-male sexual activity that are unusually high for birds, in some cases, as high as 19% of all pairs in a population.

Penguins: In early February 2004 the *New York Times* reported that Roy and Silo, a male pair of chinstrap penguins in the Central Park Zoo in New York City had successfully hatched and fostered a female chick from a fertile egg they had been given to incubate. Other penguins in New York zoos have also been reported to have formed same-sex pairs.

Zoos in Japan and Germany have also documented homosexual male penguin couples. The couples have been shown to build nests together and use a stone as a substitute for an egg. Researchers at Rikkyo University in Tokyo found 20 homosexual pairs at 16 major aquariums and zoos in Japan.

Bremerhaven Zoo in Germany attempted to encourage reproduction of endangered Humbolt penguins by importing females from Sweden and separating three male pairs, but this was unsuccessful. The zoo's director said that the relationships were "too strong" between the homosexual pairs. German gay groups protested at this attempt to break up the male-male pairs but the zoo's director was reported as saying "We don't know whether the three male pairs are really homosexual or whether they have just bonded because of a shortage of females... nobody here wants to forcibly separate homosexual couples."

A pair of male Magellanic penguins who had shared a burrow for six years at the San Francisco Zoo and raised a surrogate chick, split when the male of a pair in the next burrow died and the female sought a new mate.

Buddy and Pedro, a pair of male African Penguins, will be separated by the Toronto Zoo to mate with female penguins.

Vultures: In 1998 two male Griffon vultures named Dashik and Yehuda, at the Jerusalem Biblical Zoo, engaged in "open and energetic sex" and built a nest. The keepers provided the couple with an artificial egg, which the two parents took turns incubating; and 45 days later, the zoo replaced the egg with a baby vulture. The two male vultures raised the chick together. A few years later, however, Yehuda became interested in a female vulture that was brought into the aviary. Dashik became depressed, and was eventually moved to the zoological research garden at Tel Aviv University where he too set up a nest with a female vulture.

Two homosexual male vultures at the Allwetter Zoo in Muenster built a nest together, although they were picked on and often had their nest materials stolen by other vultures. They were eventually separated to try to promote breeding by placing one of them with female vultures, despite the protests of German homosexual groups.

Pigeons: Both male and female pigeons sometimes exhibit homosexual behaviour. As well as sexual behaviour same-sex pigeon pairs will build nests, and hens will lay (infertile) eggs and attempt to incubate them.

Some pigeons also display fetish behaviour and attempt to mate with specific inanimate objects.

Mammals

Amazon Dolphin: The Amazon River dolphin or boto has been reported to form up in bands of 3–5 individuals enjoying group sex. The groups usually comprise young males and sometimes one or two females. Sex is often performed in non-reproductive ways, using snout, flippers and genital rubbing, without regards to gender. In captivity, they have been observed to sometimes perform homosexual and heterosexual penetration of the blowhole, a hole homologous with the nostril of other mammals, making this the only known example of nasal sex in the animal kingdom. The males will sometimes also perform sex with tucuxi males, a small porpoise.

American Bison: Courtship, mounting, and full anal penetration between bulls has been noted to occur among American Bison. The Mandan nation Okipa festival concludes with a ceremonial enactment of this behaviour, to "ensure the return of the buffalo in the coming season." Also, mounting of one female by another is common among cattle.

Bonobo and Other Apes: The Bonobo, which has a matriarchal society, unusual amongst apes, is a fully bisexual species—both males and females engage in heterosexual and homosexual behaviour, being noted for female-female homosexuality in particular. About 60% of all sexual activity in this species is between two or more females. While the homosexual bonding system in Bonobos represents the highest frequency of homosexuality known in any species, homosexuality has been reported for all great apes (a group which includes humans), as well as a number of other primate species. Dutch primatologist Frans de Waal on observing and filming bonobos noted that there were two reasons to believe sexual activity is the bonobo's answer to avoiding conflict.

Anything that arouses the interest of more than one bonobo at a time, not just food, tends to result in sexual contact. If two bonobos approach a cardboard box thrown into their enclosure, they will briefly mount each other before playing with the box. Such situations lead to squabbles in most other species. But bonobos are quite tolerant, perhaps because they use sex to divert attention and to defuse tension.

Bonobo sex often occurs in aggressive contexts totally unrelated to food. A jealous male might chase another away from a female, after which the two males reunite and engage in scrotal rubbing. Or after a female hits a juvenile, the latter's mother may lunge at the aggressor, an action that is immediately followed by genital rubbing between the two adults.

Bottlenose Dolphins: Dolphins of several species engage in homosexual acts, though it is best studied in the bottlenose dolphins. Sexual encounters between females take the shape of "beak-genital propulsion", where one female insert her beak in the genital opening of the other while swimming gently forward. Between males, homosexual behaviour include rubbing of genitals against each other, which sometimes lead to the males swimming belly to belly, inserting the penis in the others genital slit and sometimes anus.

Janet Mann, Georgetown University professor of biology and psychology, argues that the strong personal behaviour among male dolphin calves is about bond formation and benefits the species in an evolutionary context. She cites studies showing that these dolphins later in life as adults are in a sense bisexual, and the male bonds forged earlier in life work together for protection as well as locating females to reproduce with. Confrontations between flocks of bottlenose dolphins and the related species Atlantic spotted dolphin will sometimes lead to cross-species homosexual behaviour between the males rather than combat.

Elephants: African and Asian males will engage in same-sex bonding and mounting. Such encounters are often associated with affectionate interactions, such as kissing, trunk intertwining, and placing trunks in each other's mouths. Male elephants, who often live apart from the general herd, often form "companionships", consisting of an older individual and one or sometimes two younger, attendant males with sexual behaviour being an important part of the social dynamic. Unlike heterosexual relations, which are always of a fleeting nature, the relationships between males may last for years. The encounters are analogous to heterosexual bouts, one male often extending his trunk along the other's back and pushing forward with his tusks to signify his intention to mount. Same-sex relations are common and frequent in both sexes, with Asiatic elephants in captivity devoting roughly 45% of sexual encounters to same-sex activity.

Giraffes

Male giraffes have been observed to engage in remarkably high frequencies of homosexual behaviour. After aggressive "necking", it is common for two male giraffes to caress and court each other, leading up to mounting and climax. Such interactions between males have been found to be more frequent than heterosexual coupling. In one study, up to 94% of observed mounting incidents took place between two males. The proportion of same sex activities varied between 30 and 75%, and at any given time one in twenty males were engaged in non-combative necking behaviour with another male. Only 1% of same-sex mounting incidents occurred between females.

Humans

Monkeys: Among monkeys, Lionel Tiger and Robin Fox conducted a study on how Depo-Provera contraceptives lead to decreased male attractiveness to females and eventually to male homosexuality. Janet E. Smith summarizes the findings as follows:

The study in the early 70s, involved a tribe of monkeys. The alpha monkey of this tribe, named Austin, chose three female monkeys to be his exclusive sexual partners. Austin had a grand time with these three female monkeys. Then the researchers injected Austin's three females with the contraceptive Depo-Provera. Austin stopped having sex with them and chose other female monkeys to be his sexual partners. Then they contracepted all of the females in the tribe. The males stopped have sex with the females and started behaving in a turbulent and confused manner.

Japanese Macaque: With the Japanese macaque, also known as the "snow monkey", same-sex relations are frequent, though rates vary between troops. Females will form "consortships" characterised by affectionate social and sexual activities. In some troops up to one quarter of the females form such bonds, which vary in duration from a few days to a few weeks. Often, strong and lasting friendships result from such pairings. Males also have same-sex relations, typically with multiple partners of the same age. Affectionate and playful activities are associated with such relations.

Lions: Both male and female lions have been seen to interact homosexually. Male lions pair-bond for a number of days and initiate homosexual activity with affectionate nuzzling and caressing, leading to mounting and thrusting. About 8% of mountings have been observed to occur with other males. Pairings between females are held to be fairly common in captivity but have not been observed in the wild.

Polecat: European polecats *Mustela putorius* were found to engage homosexually with non-sibling animals. Exclusive homosexuality with mounting and anal penetration in this solitary species serves no apparent adaptive function.

Sheep: The reason why *Ovis aries* has attracted so much attention is that some rams seem to have an exclusive homosexual orientation.

An October 2003 study by Dr. Charles E. Roselli et al. (Oregon Health and Science University) states that homosexuality in male sheep (found in 8% of rams) is associated with a region in the rams' brains which the authors call the "ovine Sexually Dimorphic Nucleus" (oSDN) which is half the size of the corresponding region in heterosexual male sheep.

Scientists found that, "The oSDN in rams that preferred females was significantly larger and contained more neurons than in male-oriented rams and ewes. In addition, the oSDN of the female-oriented rams expressed higher levels of aromatase, a substance that converts testosterone to estradiol, a form of estrogen which is believed to facilitate typical male sexual behaviours. Aromatase expression was no different between male-oriented rams and ewes."

"The dense cluster of neurons that comprise the oSDN express cytochrome P450 aromatase. Aromatase mRNA levels in the oSDN were significantly greater in female-oriented rams than in ewes, whereas male-oriented rams exhibited intermediate levels of expression." These results suggest that "...naturally occurring variations in sexual partner preferences may be related to differences in brain anatomy and its capacity for estrogen synthesis." As noted prior, given the potential unagressiveness of the male population in question, the differing aromatase levels may also have been evidence of aggression levels, not sexuality. It should also be noted that the results of this study have not been confirmed by other studies.

The Merck Manual of Veterinary Medicine appears to consider homosexuality among sheep as a routine occurrence and an issue to be dealt with as a problem of animal husbandry.

Spotted Hyena: The family structure of the Spotted Hyena is matriarchal, and dominance relationships with strong sexual elements are routinely observed between related females. Due largely to the female spotted hyena's unique urogenital system, which looks more like a penis rather than a vagina, early naturalists thought hyenas were hermaphroditic males who commonly practiced homosexuality. Early writings such as Ovid's *Metamorphoses* and the *Physiologus* suggested that the hyena continually changed its sex and nature from male to female and back again. In *Paedagogus*, Clement of Alexandria noted that the hyena (along with the hare) was "quite obsessed with sexual intercourse." Many Europeans associated the hyena with sexual deformity, prostitution, deviant sexual behaviour, and even witchcraft.

The reality behind the confusing reports is the sexually aggressive behaviour between the females, including mounting between females. Research has shown that "in contrast to most other female mammals, female *Crocuta* are male-like in appearance, larger than males, and substantially more aggressive," and they have "been masculinised without being defeminised."

Study of this unique genitalia and aggressive behaviour in the female hyena has led to the understanding that more aggressive females are better able to compete for resources, including food and mating partners. Research has shown that "elevated levels of testosterone in utero" contribute to extra aggressiveness; both males and females mount members of the same sex, who in turn are possibly acting more submissive because of lower levels of testosterone in utero.

Others

Lizards: Whiptail lizard (Teiidae genus) females have the ability to reproduce through parthenogenesis and as such males are rare and sexual breeding non-standard. Females engage in sexual behaviour to stimulate ovulation, with their behaviour following their hormonal cycles; during low levels of estrogen, these (female) lizards engage in "masculine" sexual roles. Those animals with currently high estrogen levels assume "feminine" sexual roles.

Lizards that perform the courtship ritual have greater fertility than those kept in isolation due to an increase in hormones triggered by the sexual behaviours. So, even though asexual whiptail lizards populations lack males, sexual stimuli still increase reproductive success.

From an evolutionary standpoint, these females are passing their full genetic code to all of their offspring (rather than the 50% of genes that would be passed in sexual reproduction). Certain species of gecko also reproduce by parthenogenesis.

Dragonflies

Male homosexuality has been inferred in several species of dragonflies (the order Odonata). The cloacal pinchers of male damselflies and dragonflies inflict characteristic head damage to females during sex. A survey of 11 species of damsel and dragonflies has revealed such mating damages in 20 to 80 % of the males too, indicating a fairly high occurrence of sexual coupling between males.

Fruit Flies

Male *Drosophila melanogaster* flies bearing two copies of a mutant allele in the fruitless gene court and attempt to mate exclusively with other males. The genetic basis of animal homosexuality has been studied in the fly *Drosophila melanogaster*. Here, multiple genes have been identified that can cause homosexual courtship and mating. These genes are thought to control behaviour through pheromones as well as altering the structure of the animal's brains. These studies have also investigated the influence of environment on the likelihood of flies displaying homosexual behaviour.

Bed Bugs

Male bed bugs (Cimex lectularius) are sexually attracted to any newly fed individual and this results in homosexual mounting. This occurs in heterosexual mounting by the traumatic insemination in which the male pierces the female abdomen with his needle-like penis. In homosexual mating this risks abdominal injuries as males lack the female counteradaptive spermalege structure. Males produce alarm pheromones to reduce such homosexual matings.

Sequential Hermaphroditism

Sequential hermaphroditism (called dichogamy in botany) is a type of hermaphroditism that occurs in many fish, gastropods and plants. Here, the individual is born one sex and changes sex at some point in their life. They can change from a male to female (protandry), or from female to male (protogyny). Despite which sex the organism changes to, those that change gonadal sex can have both female and male germ cells in the gonads or can change from one complete gonadal type to the other during their last life stage.

Zoology

Protandry: Protandry refers to organisms that are born male and at some point in their lifespan change sex to female. Protandrous animals include clownfish. Clownfish have a very structured society. In the *Amphiprion percula* species, there are zero to four individuals excluded from breeding and a breeding pair living in a sea anemone. Dominance is based on size, the female being the largest and the male being the second largest.

The rest of the group is made up of progressively smaller non-breeders, which have no functioning gonads. If the female dies, the male gains weight and becomes the female for that group. The largest non-breeding fish then sexually matures and becomes the male of the group.

Other examples of protandrous animals include:

- The ctenophore *Coeloplana gonoctena*. In this organism the females are bigger than the males and are only found during the summer. In contrast males are found year round.
- The flatworms *Hymanella retenuova* and *Paravortex cardii*.
- *Laevapex fuscus*, a gastropod, is described as being functionally protandric. The sperm matures in late winter and early spring, and the eggs mature in early summer, and copulation occurs only in June. This shows that males cannot reproduce until the females appear, thus why they are considered to be functionally protandric.

Protogyny

Protogyny refers to organisms that are born female and at some point in their lifespan change sex to males. Common model organisms for this type of sequential hermaphroditism are wrasses. They are one of the largest families of coral reef fish and belong to the Labridae family. Wrasses are found around the world in all marine habitats and tend to bury themselves in sand at night or when they feel threatened. In wrasses, the larger of the two fish is the male, while the smaller is the female. In most cases, females and immature have a uniform colour while the male has the terminal bicoloured phase. Large males hold territories and try to pair spawn while small to mid-size initial-phase males live with females and group spawn. In other words, both the initial and terminal phase males can breed; they differ however in the way they do it.

In the California Sheephead (*Pimelometopon pulchrum*), a type of wrasse, when the female changes to male, the ovaries degenerate and spermatogenic crypts appear in the gonads. The general structure of the gonads remains ovarian after the transformation and the sperm is transported through a series of ducts on the periphery of the gonad and oviduct. Here sex change is age dependant. For example, the California sheephead stays a female for four years before changing sex.

Other examples of protogynous organisms include:

- The isopods *Cyathura polita* and *C. carinata*
- The tanaidacean *Heterotanais oerstedi*.
- The echinoderms, *Asterina pancerii* and *A. gibbosa* are also protogynous and they brood their young.
- Protogyny sometimes occurs in the frog *Rana temporaria*, where old females sometimes change to males.

Ultimate Causes

Ghiselin proposed three models for hermaphroditism in 1969 in his paper titled "The evolution of hermaphroditism among animals". The "low-density model" states that individuals have characteristics that reduce the opportunity for mating; this model cannot be applied to sequential hermaphroditism.

The "gene dispersal model" is based on the idea that limitations on dispersal may influence population structure or genetical environment and it can be separated into two versions: the inbreeding version and the sampling-error version. This theory of gene dispersal can be applied to sequential hermaphrodites, especially the inbreeding version. The inbreeding version is based upon the fact that both protandry and protogyny help prevent inbreeding in plants and thus one can make the same assumption that in animals it works by reducing the probability of this occurring among siblings.

The sampling-error version is based on the reality that the genetical environment is influenced by genetic drift and similar phenomena in small populations. The two aspects of these hypotheses influenced by hermaphroditism, that is inbreeding and sampling-error, result in the same thing, reduction of genetic variability. In other words being a hermaphrodite would increase genetic variability and thus be considered advantageous to the organism. This theory of gene dispersal can be applied to sequential hermaphrodites, especially the inbreeding version. Lastly, the "size-advantage model" states that reproductive functions are carried out better if the individual is a certain size/age. Assuming that the reproductive functions of one sex are better performed at a certain size, then an organism would assume the sex that its size allows to perform the best. This would increase its reproductive potential and fitness. For example, eggs are larger than sperm, thus if you are a big you are able to make more eggs so being female when big is advantageous, however the size advantage relationship is really not as simple as the example just mentioned, but it allows for a better understanding of it.

In most ectotherms body size and female fecundity are positively correlated. This supports Ghiselin's size-advantage model, which is still widely accepted today. Kazancioglu and Alonzo (2010) performed the first comparative analysis of sex change in Labridae. Their analysis supports the size-advantage model by Ghiselin and suggest that sequential hermaphroditism is correlated to the size-advantage. They determined that dioecy was less likely to occur when the size advantage is stronger than other advantages

Warner suggests that selection for protandry may occur in populations where female fecundity is augmented with age and individuals mate randomly. Selection for protogyny may occur where there are traits in the population that depress male fecundity at early ages (territoriality, mate selection or inexperience) and when female fecundity is decreased with age, the latter seems to be rare in the field. An example of territoriality favouring protogyny occurs when there is a need to protect their habitat and being a large male is advantageous for this purpose. In the mating aspect, a large male has a higher chance of mating, while this has no effect on the female mating fitness. Thus, he suggests that female fecundity has more impact on sequential hermaphroditism that the age structures of the population.

The size-advantage model predicts that sex change would only be absent if the relationship between size/age with reproductive potential is identical in both sexes.

With this prediction one would assume that hermaphroditism is very common, but this is not the case. Sequential hermaphroditism is very rare and according to scientists this is due to some cost that decreases fitness in sex changers as opposed to those who don't change sex. Kazanciglu and Alonzo confirmed this in 2009.

They found that the costs of changing sex only favoured dioecy when the cost was very large but that some groups favoured hermaphroditism. This indicates that the cost of sex change does not explain the rarity of sequential hermaphroditism by itself.

Proximate Causes

Many studies have focused on the proximate causes of sequential hermaphroditism. The role of aromatase has been widely studied in this area. Aromatase is an enzyme that controls the androgen/estrogen ratio in animals by catalysing the conversion of testosterone into oestradiol, which is irreversible. It has been discovered that the aromatase pathway mediates sex change in both directions. Many studies also involve understanding the effect of aromatase inhibitors on sex change. One such study was performed by Kobayashi et al. In their study they tested the role of estrogens in male three-spot wrasses (*Halichoeres trimaculatus*). They discovered that fish treated with aromatase inhibitors showed decreased gonodal weight, plasma estrogen level and spermatogonial proliferation in the testis as well as increased androgen levels. Their results suggest that estrogens are important in the regulation of spermatogenesis in this protogynous hermaphrodite.

Animals Displaying Homosexual Behaviour

Animals (birds, mammals, insects, reptiles, fish etc.) for which there is documented evidence of homosexual or transgender behaviour of one or more of the following kinds: sex, courtship, affection, pair bonding, or parenting, as noted in researcher and author Bruce Bagemihl's 1999 book *Biological Exuberance: Animal Homosexuality and Natural Diversity*.

Bagemihl writes that the presence of same-sex sexual behaviour was not 'officially' observed on a large scale until the 1990s due to possible observer bias caused by social attitudes towards LGBT people making the homosexual theme taboo. Bagemihl devotes three chapters; *Two Hundred Years at Looking at Homosexual Wildlife, Explaining (Away) Animal Homosexuality* and *Not For Breeding Only* in his 1999 book *Biological Exuberance* to the "documentation of systematic prejudices" where he notes "the *present ignorance* of biology lies precisely in its single-minded attempt to find reproductive (or other) "explanations" for homosexuality, transgender, and non-procreative and alternative heterosexualities.

Petter Bøckman, academic adviser for the *Against Nature?* Exhibit stated "Many researchers have described homosexuality as something altogether different from sex. They must realise that animals can have sex with who they will, when they will and without consideration to a researcher's ethical principles". Homosexual behaviour is found amongst social birds and mammals, particularly the sea mammals and the primates. Animal sexual behaviour takes many different forms, even within the same species and the motivations for and implications of their behaviours have yet to be fully understood. Bagemihl's research shows that homosexual behaviour, not necessarily sex, has been observed in close to 1500 species, ranging from primates to gut worms, and is well documented for 500 of them.

Homosexuality in animals is seen as controversial by social conservatives because it asserts the naturalness of homosexuality in humans, while others counter that it has no implications and is nonsensical to equate animal behaviour to morality. Animal preference and motivation is always inferred from behaviour. Thus homosexual behaviour has been given a number of terms over the years. The correct usage of the term *homosexual* is that an animal *exhibits homosexual behaviour*, however this article conforms to the usage by modern research applying the term *homosexuality* to all sexual behaviour (copulation, genital stimulation, mating games and sexual display behaviour) between animals of the same sex.

The all-female Whiptail lizard species *Cnemidophorus neomexicanus* (centre), which reproduces via parthenogenesis, is shown flanked by two sexual species having males, *C. inornatus* (left) and *C. tigris* (right). Research has shown that simulated mating behaviour increases fertility for *Cnemidophorus neomexicanus*. One female lies on top of another, playing the role of the male, the lizard that was on bottom has larger eggs. The lizards switch off this role each mating season.

Male homosexuality has been inferred in several species of dragonflies. A survey of damsel and dragonflies reveals characteristic cloacal pincher mating damage in 20–80 percent of the males, indicating a fairly high occurrence of sexual coupling between males.

Mammals

Selected mammals from the full list:

- Bison
- Brown Bear
- Brown Rat
- Caribou
- Cat (domestic)

- Cattle (domestic)
- Cheetah
- Chimpanzee
- Common Dolphin
- Common Marmoset
- Dog
- Elephant
- Fox
- Giraffe
  Goat
- Horse (domestic)
- Human
- Koala
- Lion
- Orca
- Raccoon.

Birds

Selected birds from the full list:
- Barn Owl
- Chicken
- Common Gull
- Emu

- King Penguin
- Mallard
- Raven
- Seagull.

Fish

- Amazon molly
- Blackstripe topminnow
- Bluegill Sunfish
- Char
- Grayling

- European Bitterling
- Green swordtail
- Guiana leaffish
- Houting Whitefish
- Jewel Fish
- Least Darter (*Microperca punctulata*)
- Mouthbreeding Fish sp.
- Salmon spp.
- Southern platyfish
- Ten-spined stickleback
- Three-spined stickleback.

Reptiles

- Anole sp.
- Bearded Dragon
- Broad-headed Skink
- Checkered Whiptail Lizard
- Chihuahuan Spotted Whiptail Lizard
- Common Ameiva
- Common Garter Snake
- Cuban Green Anole
- Desert Grassland Whiptail Lizard
- Desert Tortoise
- Fence Lizard
- Five-lined Skink
- Gopher (Pine) Snake
- Green Anole
- Inagua Curlytail Lizard
- Jamaican Giant Anole
- Laredo Striped Whiptail Lizard
- Largehead Anole
- Mourning Gecko
- Plateau Striped Whiptail Lizard
- Red Diamond Rattlesnake

- Red-tailed Skink
- Side-blotched Lizard
- Speckled Rattlesnake
- Water Moccasin
- Western rattlesnake (Crotalus viridis)
- Western Banded Gecko
- Whiptail Lizard spp.
- Wood Turtle.

Amphibians

- Appalachian Woodland Salamander
- Black-spotted Frog
- Mountain Dusky Salamander
- Tengger Desert Toad.

Insects

- Alfalfa Weevil
- Australian Parasitic Wasp sp.
- Bean Weevil sp.
- Bedbug and other Bug spp.
- Blister Beetle spp.
- Blowfly
- Broadwinged Damselfly sp.
- Cabbage (Small) White (Butterfly)
- Checkerspot Butterfly
- Club-tailed Dragonfly spp.
- Cockroach spp.
- Common Skimmer Dragonfly spp.
- Creeping Water Bug sp.
- Cutworm
- Digger Bee
- Dragonfly spp.
- Eastern Giant Ichneumon Wasp
- Eucalyptus Longhorned Borer

- Field Cricket sp.
- Flour Beetle
- Fruit Fly spp.
- Glasswing Butterfly
- Grape Berry Moth
- Grape Borer
- Green Lacewing
- Hen Flea
- House Fly
- Ichneumon wasp sp.
- Japanese Scarab Beetle
- Larch Bud Moth
- Large Milkweed Bug
- Large White
- Long-legged Fly spp.
- Mazarine Blue
- Mediterranean Fruit Fly
- Mexican White (butterfly)
- Midge sp.
- Migratory Locust
- Monarch Butterfly
- Narrow-winged Damselfly spp.
- Parsnip Leaf Miner
- Pomace fly
- Queen Butterfly
- Red Ant sp.
- Red Flour Beetle
- Reindeer Warble Fly (Hypoderma tarandi)
- Rose Chafer
- Rove Beetle spp.
- Scarab Beetle (Melolonthine)
- Screwworm Fly
- Silkworm Moth

- Southeastern Blueberry Bee
- Southern Green Stink Bug
- Southern Masked Chafer
- Southern One-Year Canegrub
- Spreadwinged Damselfly spp.
- Spruce Budworm Moth
- Stable Fly sp.
- Stag Beetle spp.
- Tsetse Fly
- Water Boatman Bug
- Water Strider spp.

Other Invertebrates

- Blood-Fluke
- Box Crab
- Harvest Spider sp.
- Hawaiian Orb-Weaver (spider)
- Incirrate Octopus spp.
- Jumping Spider sp.
- Mite sp.
- Spiny-Headed Worm.

Seabird Breeding Behaviour

The term seabird is used for many families of birds in several orders that spend the majority of their lives at sea. Seabirds make up some, if not all, of the families in the following orders: Procellariiformes, Sphenisciformes, Pelecaniformes, and Charadriiformes. Many seabirds remain at sea for several consecutive years at a time, without ever seeing land. Breeding is the central purpose for seabirds to visit land.

The breeding period (courtship, copulation, and chick-rearing) is usually extremely protracted in many seabirds and may last over a year in some of the larger albatrosses ; this is in stark contrast with passerine birds. Seabirds nest in single or mixed-species colonies of varying densities, mainly on offshore islands devoid of terrestrial predators . However, seabirds exhibit many unusual breeding behaviours during all stages of the reproductive cycle that are not extensively reported outside of the primary scientific literature.

Courtship Stage

The courtship stage of breeding is when pair bonds are formed and occurs before copulation and occasionally continues through the copulatory and chick-rearing stages of the breeding phenology. The sequence and variety of courting behaviours vary widely among species, but they typically begin with territorial defence, followed by mate-attraction displays, and selection of a nest site . Seabirds are long-lived, socially monogamous, birds that usually mate for life. This makes selecting a mate extremely important with lifelong implications for the reproductive success of both individuals in the pair.

Mating Dances

Seabirds are one of the only avian families that include ritualised dances in their courtship. These dances are complex and can include displays and vocalisation that vary greatly between families and orders. Albatrosses are well known for their intricate mating dances. All species of albatross have some form of ritualised dance, with many species displaying very similar forms. Albatrosses' complex visual and vocal dances are considered some of the most developed mating displays in any long-lived animal . Both members of the pair use these dances as a proxy for mate quality and it is believed to be a very important aspect of mate choice in this family .

For Black-footed (*Phoebastria nigripes*) and Laysan Albatrosses (*P. immutabilis*) there are ten described parts to their mating dance which can be given in various sequences . Several parts include "billing" where one individual gently touches the others bill and "sky pointing" where the bird rises on the tips of its toes, stretches its neck and points its bill upward.

In the Wandering Albatross (*Diomedea exulans*), sky pointing is accompanied with "sky calling" where the displaying individual spreads its wings, revealing his massive 12 foot wingspan while pointing and vocalising skyward . The mating dance may last for several minutes. It has been noted that many albatross species dance upon reuniting with their partner every year; however, for Waved Albatross (*P. irrorata*), the dance is longer and more involved in new pairs, or in pairs that failed to breed the previous season .

Boobies are another group of seabirds known for their mating displays. Brown (*Sula leucogaster*), Red-footed (*S. sula*) and Blue-footed Boobies (*S. nebouxii*) have at least nine described parts to their mating display . Sky pointing in boobies is similar to albatrosses; in the Brown Booby, sky pointing is described as a display where the male throws his head backwards, stretches his neck out, and usually gives a whistling vocalisation .

Parading is a well-known display in boobies as well; in this display, one individual in the pair - usually the male - walks upright, with his tail erect, swaying in an exaggerated manner from side to side while taking small steps . In Blue-footed and Red-footed Boobies, parading also includes lifting their brightly coloured feet to flaunt to their partner.

Frigatebirds are known for their unusual displays and breeding system. Unlike other seabirds, frigatebirds have a lek-breeding system where displaying males aggregate in groups of up to 30 individuals with prospecting females flying overhead . However, unlike classic leks, the pair then builds a nest on the male's display site. The male then participates fully in nest defence, incubation, and chick-rearing . The main display that male frigatebirds use to attract females is a "gular presentation" where the male inflates his bright red throat pouch, points his head upwards and opens his wings . Interestingly, it has been shown experimentally that there is no correlation between energy expended by males during courtship display and mate selection by females .

Courtship Feeding

Once the pair bond is formed, courtship feeding occurs in some species. Courtship feeding is when one member of the pair presents the other with food in a ritualised way. Often the male feeds the female, but in certain species where the sex roles are reversed, the female may feed the male. Several reasons proposed as to why courtship feeding occurs is: 1) to help strengthen the pair bond 2) to reduce aggression between males and females and 3) to provide additional nutrition to the females during the egg-laying stage.

Courtship feeding is seen in many gull and tern species. In Common Terns (*Sterna hirundo*), courtship feeding begins right at the start of pair formation with male terns carrying a fish around the breeding colony, displaying it to prospective mates. The direct benefits hypothesis (where the female obtains some immediate benefit for copulating with the male, food in this case) may explain why courtship feeding has evolved; however, this theory has recently been disputed with the suggestion that the rate of courtship feeding is a way for females to determine the quality of their mate through the handicap principle.

Same-sex Pairing

Homosexual behaviour has been well documented in over 500 species of non-human animals ranging from insects to lizards to mammals (reviewed in:). In birds, same-sex pairing has been shown in many families of non-passerines including vultures, ducks, and pigeons . There is also a remarkably high incidence of homosexual behaviour in seabirds. Here, homosexual behaviour refers to same-sex pair-formation and chick-rearing, not to same-sex copulation, for which there are very few documented examples. Almost all the examples of same-sex pairing in seabirds are of female-female pairs. Furthermore, this phenomenon doesn't seem to be phylogenetically constrained to any specific order or family of seabirds.

There are many examples of homosexual behaviour in wild gulls. In Herring Gull (*Larus argentatus*) populations nesting on the Great Lakes, Fitch (1980) reported a low, yet consistent prevalence of female-female pairs. It appears that female-female Herring Gull pairs are more common in colonies with a female-biased operational sex ratio (OSR) and occasionally these homosexual pairs will remain stable for several breeding seasons . In Western Gulls (*Larus occidentalis*), female-female pairs are often associated with supernormal clutches (clutches of 4-6 eggs; a normal clutch for *Larus* gull species is 2-3 eggs) and these clutches are usually infertile . Female-female pairs have also been widely reported in wild populations of Ring-billed Gulls (*Larus delawarensis*). Studies of Ring-billed Gulls has shown that same sex pairs are rare (<1% of pairs in a colony) but consistent interannually and that they also lay supernormal clutches at a significantly higher rate than do heterosexual pairs. It has also been shown that these clutches of female-female pairs have significantly lower hatching and fledging success than heterosexual pairs . There is even one example of an unusual mixed female-female pair of two gull species, the Caspian (*Larus cachinnans*) and Yellow-legged Gull (*Larus michahellis*) .

Female-female pairing has also been documented and studied in several tern species including Whiskered (*Chlidonias hybyida*), Roseate (*Sterna dougalii*) and Caspian Terns (*Hydroprogne caspia*) with similar attributes (supernormal clutches and reduced hatching/fledging success as compared to heterosexual pairs) to same-sex gull pairs . Also in the order Charadriiformes (family Chionididae), there has been one reported occurrence of a female-female pair in the Black-faced Sheathbill (*Chionis minor*), but eggs in the clutch proved to be inviable.

Same-sex pairing has also been shown in several families of true seabirds including the petrels and shearwaters. Antarctic Petrels (*Thalassoica antarctica*) have been shown to form female-female pairs in colonies where there is a surplus of females; it is hypothesized that "pairing" with another female may be a favourable strategy for some females because it allows them to become established in the colony. The experience with a site gained through forming a female-female pair may greatly improve the chances of future successful breeding for the non-genetic parent, which explains why it might be worth the short-term cost of raising another bird's offspring . In another member of this family, the Cory's Shearwater (*Calonectris diomedea*), same-sex pairing was recently discovered for the first time in a burrow-nesting seabird . This study proposed that similar factors cause female-female pairs to form in burrow-nesting seabirds as in surface-nesting seabirds (a female-biased OSR), and that female-female pairing in burrow-nesting seabirds might have remained undetected for so long due to the secretive nature of these animals.

In albatrosses, female-female pairing has recently received major press coverage. Last year, when a Southern Royal Albatross (*D. epomophora*) couple hatched a chick in New Zealand, it represented the first record of a successful same-sex pair in this species . In a landmark study by Young et al. (2008), she reported over 30% of Laysan Albatrosses in a colony in Oahu, Hawaii were same sex pairs. Even though these female-female pairs had less reproductive success than heterosexual pairs, it was better than not breeding at all. Young et al. (2008) also cited a female-biased OSR as the primary reason for such a high proportion of same-sex pairs. Additionally, an unsuccessful female-female pair of highly endangered of Short-tailed Albatrosses (*P. albatrus*) was recently documented on Kure Atoll, Hawaii.

Penguins represent the only known examples of male-male pairings in seabirds. On the Otago Peninsula of New Zealand, two-male Yellow-eyed Penguins (*Megadyptes antipodes*) were recently reported incubating an egg. In captivity, Chinstrap (*Pygoscelis antarcticus*), Humboldt (*Spheniscus humboldti*), Magellanic (*S. magellanicus*), and African Black-footed Penguins (*S. demersus*) have all been documented to form male-male pairs.

Copulatory Stage

In seabirds, the copulatory stage usually occurs after, and occasionally concurrently, with the formation of the pair bond.

Copulation occurs mainly on land at the breeding colony. Usually the pair copulates several times, even in orders that lay only one egg per-clutch. These additional copulations are thought of as a mechanism to strengthen the pair bond . This is important for strongly monogamous, long-lived organisms and is especially important in seabirds that spend most of the non-breeding season apart on the open ocean.

Extra-pair Copulation/Fertilization/Paternity

Birds are one of the only major taxa where monogamy is the dominant mating system . Prior to the advent of genetic techniques, it was assumed that the majority of monogamous birds remained faithful to their partners . However, it is now known that extra-pair copulations (EPCs), extra-pair fertilizations (EPFs), and extra-pair paternity (the raising of another's offspring, EPP) are actually quite common in a variety of avian orders and families . Roughly 70% of birds that used to be considered genetically monogamous actually engage in EPCs and raise extra-pair young (reviewed by:). Furthermore, it has been proposed that birds that nest in high densities, as seabirds do in breeding colonies, have higher rates of EPCs and EPFs than birds that do not nest colonially. Despite this, Westneat and Sherman (1997) found no significant correlation between nesting density and EPFs in a meta-analysis. Many seabird species raise only one chick per breeding season, which would make the prevalence of EPFs and EPP in seabirds surprising due to the fact that the male's entire breeding success for a year is dependent on the lone egg/chick he is raising to be his genetic offspring. Moreover, all seabirds have obligate biparental care, so it would be evolutionarily costly for the male to spend months of effort raising a chick that is not his genetic offspring.

In line with this prediction, many studies of seabirds have revealed no EPCs or EPFs. Several genetic studies of Storm-Petrels, show no evidence of EPCs or EPFs, which is not surprising considering these are burrow-nesting seabirds that lay only one egg per year. Dovekie (*Alle alle*), a surface nesting alcid that raises one chick per year, has shown no EPFs or EPP . Nazca Boobies (*Sula granti*) have been well studied at breeding colonies in the Galapagos for decades and also show no evidence of EPCs or EPFs; also not a surprising result since they only have one surviving offspring per year . Chinstrap Penguins, which raise two chicks annually, have also shown no EPCs or EPFs.

Contrary to these empirical results, there has been a multitude of studies where EPCs or EPFs have been found in seabirds. Perhaps the most surprising EPFs have been found in the Procellariid family because all members of this family only lay one egg per year and some do not even breed every year . Waved Albatross show high rates (up to 25% of offspring are extra-pair young) of both EPCs and EPFs and behavioural observations have shown that many of the EPCs are forced by the extra-pair male. Studies of Wandering Albatross have shown over 10% of chicks are extra-pair young; an extremely surprising result since adult Wandering Albatrosses only breed once every other year at most. Similar results were seen in Black-browed (*Thalassarche melanophris*) and Grey-headed Albatrosses (*T. chrysostoma*) nesting on South Georgia Island. EPFs have also been shown in Antarctic Petrels. Perhaps the most unexpected result was when EPCs and EPFs were documented in a burrow-nesting Procellarid, the Short-tailed Shearwater (*Puffinus tenuirostris*), because the female would have to leave her burrow to solicit an EPC .

EPCs and EPFs have also been demonstrated to occur in other families of seabirds. In contrast to the results found in genetic studies of Dovekie, EPP has been shown in several species of alcid including Common Murre (*Uria aalge*) and Razorbill (*Alca torda*), both of which raise only one chick per year. Interestingly, it has been shown that female Razorbills can determine whether or not an EPC leads to an EPF and only accept extra-pair sperm when it gives them a fitness advantage over their current mate . A low-rate of EPP has also been shown in the sexually dimorphic Great Frigatebird (*Fregata minor*).

Inbreeding

Mating with related individuals is rare in naturally occurring populations of birds due to the production of lower quality offspring suffering from the genetic effects of inbreeding depression. Seabirds have an inherently high risk of inbreeding because most are natally philopatric, and many are highly endangered with some species' entire populations breeding on one small island . Despite this, inbreeding was observed no more than expected by random chance in the Wandering Albatross and the critically endangered Amsterdam Albatross (*D. amsterdamensis*). In contrast, some studies of seabirds have shown evidence of inbreeding. Huyvaert and Parker (2010) detected low frequencies of inbreeding in Waved Albatrosses and genetic similarity was negatively related to EPFs, which is an unusual result that does not support the inbreeding avoidance hypothesis. Close inbreeding was observed at low frequencies in the Cory's Shearwater where two mother-son pairs were reported.

Chick-rearing Stage

Chick-rearing is the most crucial stage of the reproductive cycle in determining final reproductive success during a breeding season. Chick-rearing includes brooding, feeding, defending, and in some cases, teaching the chick skills it will need to know to survive independently. Chick-rearing can be totally absent in some birds (the Brush-Turkeys of southeast Asia), to a couple weeks long in many passerines, to several months long in larger birds . Seabirds have the longest chick-rearing stage of any bird on earth . It is not unusual for many seabirds to spend 3–4 months raising their chicks until they are able to fledge and forage independently. In the great albatrosses, chick-rearing can take over 9 months . It is because of this extremely protracted chick-rearing stage that many of the larger procellariiform seabirds can breed only once every other year.

Siblicide

Siblicide, the death of an individual due to the actions of members of its own clutch, is seen in several avian orders including egrets and kingfishers, some raptors, and grackles. In most of these examples, siblicide is facultative (i.e. not obligate) and only occurs when there is a shortage of food.

However in some seabirds, siblicide proves to be obligate and occurs no matter how productive the breeding season is. The Nazca Booby is one species that practices obligate siblicide . The parents lay two eggs, several days apart. The second egg laid is seen as fertility insurance if the first egg is inviable. If both eggs hatch, the elder chick will push its sibling out of the nest area, leaving it to die of thirst or cold. The parent booby will not intervene and the younger chick will inevitably die. Research has shown that high hormone levels in Nazca Booby chicks are responsible for inciting their murderous behaviour. Facultative siblicide is seen in the closely related Blue-footed Booby. Unlike the Nazca Booby, Blue-footed Boobies chicks only perform siblicide when food is scarce. Furthermore, the parents actually try to suppress the siblicidal behaviour, rather than ignoring or encouraging it.

Pair-Splitting

Divorce: For most seabirds if breeding is successful they will continue breeding with the same partner year after year until one member of the pair dies or doesn't return to the breeding colony. However, these pair bonds occasionally dissolve or are forced apart while both members of the pair are still alive, a process known as divorce. Reasons for divorce in seabirds are wide ranging and include asynchronous arrival of mates to the breeding colony, declining reproductive success of the pair, and competition for mates. Coercive divorce is seen in the Nazca Booby and Common Murre where one member of the pair actively deserts the other or where an intruder enters and forcibly splits the breeding pair to form a new pair.

Divorce is relatively common in gulls and their relatives (reviewed by:); in one study, Black-legged Kittiwakes (*Rissa tridactyla*) proved to be more faithful to their nesting site than their partner. Divorce rates are surprisingly high (>80% of pairs annually) in King (*Aptenodytes patagonicus*) and Emperor Penguins (*A. forsteri*). Asynchronous arrival of mates at the breeding colony is cited as the main reason for this because these penguins have extreme time constraints on their breeding. In Great Skuas (*Stercorarius skua*) divorce occurs annually, but at low frequencies (6-7% of pairs annually) and death is responsible for approximately three times more pair interruptions than divorce. Divorce is uncommon in procellariiforms and usually only occurs after several years of breeding failure. However, one study of Short-tailed Shearwaters observed the divorce rate in a colony to be as high as 16% annually.

Cross Species Sex

While it is commonly believed that animal sexuality is instinctive and thus somewhat mechanistic, research regularly records that many animals are sexual opportunists, partaking in sexual relations with individuals of visibly distinct species. This is more visible in domesticated species and animals in captivity, as domestication commonly selects for increased breeding rate (and so an accelerated breeding cycle has commonly arisen in domesticated species over the centuries), and also because these species are more easily observed by humans. Nevertheless, animals have been observed in the wild to attempt sexual activity with other species or indeed inanimate objects.

In the wild, where observation is harder, genetic studies have shown a "large number" of inter-species hybrids, and other investigations describe productive and non-productive inter-species mating as a "natural occurrence". Recent genetic evidence strongly suggesting this has occurred even within the history of the human species, and that early humans often had sexual activity with other primate species.

Hybrid offspring can result from two organisms of distinct but closely related parent species, although the resulting offspring is not always fertile.

Due to the difficulties of observation, interspecies sex of this kind between two top-level predators, occurring in the wild, was only conclusively documented with the finding of a grizzly-polar bear hybrid in April 2006. Again, as with lions and tigers, the two species would normally not share enough common territory to provide adequate opportunity for much cross-species sexual activity.

Animal sexual advances on, and attempted interactions with, humans and other species, have been documented by ethologists such as Kohler, Gerald Durrell and Desmond Morris, as well as authoritative researchers such as Birute Galdikas who studied orangutans in Borneo. Philosopher and animal welfare activist Peter Singer reports:

While walking through the camp with Galdikas, my informant was suddenly seized by a large male orangutan, his intentions made obvious by his erect penis. Fighting off so powerful an animal was not an option, but Galdikas called to her companion not to be concerned, because the orangutan would not harm her, and adding, as further reassurance, that 'they have a very small penis' ... though the orangutan lost interest before penetration took place.

Prostitution

In some penguin species, the females, even when in a committed relationship, will exchange sexual favours with strange males for the pebbles they need to build their nests. Prostitution was also observed among chimpanzees, who trade food for sex.

Operant Conditioning

Operant conditioning is a form of psychological learning during which an individual modifies the occurrence and form of its own behaviour due to the association of the behaviour with a stimulus. Operant conditioning is distinguished from classical conditioning (also called respondent conditioning) in that operant conditioning deals with the modification of "voluntary behaviour" or operant behaviour. Operant behaviour "operates" on the environment and is maintained by its consequences, while classical conditioning deals with the conditioning of reflexive (reflex) behaviours which are elicited by antecedent conditions. Behaviours conditioned via a classical conditioning procedure are not maintained by consequences.

Reinforcement, Punishment, and Extinction

Reinforcement and punishment, the core tools of operant conditioning, are either positive (delivered following a response), or negative (withdrawn following a response). This creates a total of four basic consequences, with the addition of a fifth procedure known as extinction (i.e. no change in consequences following a response).

It is important to note that actors are not spoken of as being reinforced, punished, or extinguished; it is the actions that are reinforced, punished, or extinguished. Additionally, reinforcement, punishment, and extinction are not terms whose use is restricted to the laboratory. Naturally occurring consequences can also be said to reinforce, punish, or extinguish behaviour and are not always delivered by people.

- Reinforcement is a consequence that causes a behaviour to occur with greater frequency.
- Punishment is a consequence that causes a behaviour to occur with less frequency.
- Extinction is the lack of any consequence following a behaviour. When a behaviour is inconsequential (i.e., producing neither favourable nor unfavourable consequences) it will occur with less frequency. When a previously reinforced behaviour is no longer reinforced with either positive or negative reinforcement, it leads to a decline in that behaviour.

Four Contexts of Operant Conditioning

Here the terms *positive* and *negative* are not used in their popular sense, but rather: *positive* refers to addition, and *negative* refers to subtraction.

What is added or subtracted may be either reinforcement or punishment. Hence *positive punishment* is sometimes a confusing term, as it denotes the "addition" of a stimulus or increase in the intensity of a stimulus that is aversive (such as spanking or an electric shock). The four procedures are:

1. Positive reinforcement (Reinforcement): occurs when a behaviour (response) is followed by a stimulus that is appetitive or rewarding, increasing the frequency of that behaviour. In the Skinner box experiment, a stimulus such as food or sugar solution can be delivered when the rat engages in a target behaviour, such as pressing a lever.

2. Negative reinforcement (Escape): occurs when a behaviour (response) is followed by the removal of an aversive stimulus, thereby increasing that behaviour's frequency. In the Skinner box experiment, negative reinforcement can be a loud noise continuously sounding inside the rat's cage until it engages in the target behaviour, such as pressing a lever, upon which the loud noise is removed.

3. Positive punishment (Punishment) (also called "Punishment by contingent stimulation"): occurs when a behaviour (response) is followed by a stimulus, such as introducing a shock or loud noise, resulting in a decrease in that behaviour.

4. Negative punishment (Penalty) (also called "Punishment by contingent withdrawal"): occurs when a behaviour (response) is followed by the removal of a stimulus, such as taking away a child's toy following an undesired behaviour, resulting in a decrease in that behaviour.

Also:

• Avoidance learning is a type of learning in which a certain behaviour results in the cessation of an aversive stimulus. For example, performing the behaviour of shielding one's eyes when in the sunlight (or going outdoors) will help avoid the aversive stimulation of having light in one's eyes.

• Extinction occurs when a behaviour (response) that had previously been reinforced is no longer effective. In the Skinner box experiment, this is the rat pushing the lever and being rewarded with a food pellet several times, and then pushing the lever again and never receiving a food pellet again. Eventually the rat would cease pushing the lever.

• Noncontingent reinforcement refers to delivery of reinforcing stimuli regardless of the organism's (aberrant) behaviour. The idea is that the target behaviour decreases because it is no longer necessary to receive the reinforcement. This typically entails time-based delivery of stimuli identified as maintaining aberrant behaviour, which serves to decrease the rate of the target behaviour. As no measured behaviour is identified as being strengthened, there is controversy surrounding the use of the term noncontingent "reinforcement".

- Shaping is a form of operant conditioning in which the increasingly accurate approximations of a desired response are reinforced.

- Chaining is an instructional procedure which involves reinforcing individual responses occurring in a sequence to form a complex behaviour.

Thorndike's Law of Effect

Operant conditioning, sometimes called *instrumental conditioning* or *instrumental learning*, was first extensively studied by Edward L. Thorndike (1874–1949), who observed the behaviour of cats trying to escape from home-made puzzle boxes. When first constrained in the boxes, the cats took a long time to escape. With experience, ineffective responses occurred less frequently and successful responses occurred more frequently, enabling the cats to escape in less time over successive trials. In his law of effect, Thorndike theorised that successful responses, those producing *satisfying* consequences, were "stamped in" by the experience and thus occurred more frequently. Unsuccessful responses, those producing *annoying* consequences, were *stamped out* and subsequently occurred less frequently. In short, some consequences *strengthened* behaviour and some consequences *weakened* behaviour. Thorndike produced the first known learning curves through this procedure.

B.F. Skinner (1904–1990) formulated a more detailed analysis of operant conditioning based on reinforcement, punishment, and extinction. Following the ideas of Ernst Mach, Skinner rejected Thorndike's mediating structures required by "satisfaction" and constructed a new conceptualization of behaviour without any such references. So, while experimenting with some homemade feeding mechanisms, Skinner invented the operant conditioning chamber which allowed him to measure rate of response as a key dependent variable using a cumulative record of lever presses or key pecks.

Biological Correlates of Operant Conditioning

The first scientific studies identifying neurons that responded in ways that suggested they encode for conditioned stimuli came from work by Mahlon deLong and by R.T. "Rusty" Richardson. They showed that nucleus basalis neurons, which release acetylcholine broadly throughout the cerebral cortex, are activated shortly after a conditioned stimulus, or after a primary reward if no conditioned stimulus exists. These neurons are equally active for positive and negative reinforcers, and have been demonstrated to cause plasticity in many cortical regions. Evidence also exists that dopamine is activated at similar times. There is considerable evidence that dopamine participates in both reinforcement and aversive learning. Dopamine pathways project much more densely onto frontal cortex regions. Cholinergic projections, in contrast, are dense even in the posterior cortical regions like the primary visual cortex. A study of patients with Parkinson's disease, a condition attributed to the insufficient action of dopamine, further illustrates the role of dopamine in positive reinforcement. It showed that while off their medication, patients learned more readily with aversive consequences than with positive reinforcement. Patients who were on their medication showed the opposite to be the case, positive reinforcement proving to be the more effective form of learning when the action of dopamine is high.

Factors that Alter the Effectiveness of Consequences

When using consequences to modify a response, the effectiveness of a consequence can be increased or decreased by various factors. These factors can apply to either reinforcing or punishing consequences.

1. Satiation/Deprivation: The effectiveness of a consequence will be reduced if the individual's "appetite" for that source of stimulation has been satisfied. Inversely, the effectiveness of a consequence will increase as the individual becomes deprived of that stimulus. If someone is not hungry, food will not be an effective reinforcer for behaviour. Satiation is generally only a potential problem with primary reinforcers, those that do not need to be learned such as food and water.

2. Immediacy: After a response, how immediately a consequence is then felt determines the effectiveness of the consequence. More immediate feedback will be more effective than less immediate feedback. If someone's license plate is caught by a traffic camera for speeding and they receive a speeding ticket in the mail a week later, this consequence will not be very effective against speeding. But if someone is speeding and is caught in the act by an officer who pulls them over, then their speeding behaviour is more likely to be affected.

3. Contingency: If a consequence does not contingently (reliably, or consistently) follow the target response, its effectiveness upon the response is reduced. But if a consequence follows the response consistently after successive instances, its ability to

modify the response is increased. The schedule of reinforcement, when consistent, leads to faster learning. When the schedule is variable the learning is slower. Extinction is more difficult when learning occurs during intermittent reinforcement and more easily extinguished when learning occurs during a highly consistent schedule.

4. Size: This is a "cost-benefit" determinant of whether a consequence will be effective. If the size, or amount, of the consequence is large enough to be worth the effort, the consequence will be more effective upon the behaviour. An unusually large lottery jackpot, for example, might be enough to get someone to buy a one-dollar lottery ticket (or even buying multiple tickets). But if a lottery jackpot is small, the same person might not feel it to be worth the effort of driving out and finding a place to buy a ticket. In this example, it's also useful to note that "effort" is a punishing consequence. How these opposing expected consequences (reinforcing and punishing) balance out will determine whether the behaviour is performed or not.

Most of these factors exist for biological reasons. The biological purpose of the Principle of Satiation is to maintain the organism's homeostasis. When an organism has been deprived of sugar, for example, the effectiveness of the taste of sugar as a reinforcer is high. However, as the organism reaches or exceeds their optimum blood-sugar levels, the taste of sugar becomes less effective, perhaps even aversive.

The Principles of Immediacy and Contingency exist for neurochemical reasons. When an organism experiences a reinforcing stimulus, dopamine pathways in the brain are activated. This network of pathways "releases a short pulse of dopamine onto many dendrites, thus broadcasting a rather global reinforcement signal to postsynaptic neurons." This results in the plasticity of these synapses allowing recently activated synapses to increase their sensitivity to efferent signals, hence increasing the probability of occurrence for the recent responses preceding the reinforcement. These responses are, statistically, the most likely to have been the behaviour responsible for successfully achieving reinforcement. But when the application of reinforcement is either less immediate or less contingent (less consistent), the ability of dopamine to act upon the appropriate synapses is reduced.

Operant Variability

Operant variability is what allows a response to adapt to new situations. Operant behaviour is distinguished from reflexes in that its response topography (the form of the response) is subject to slight variations from one performance to another. These slight variations can include small differences in the specific motions involved, differences in the amount of force applied, and small changes in the timing of the response. If a subject's history of reinforcement is consistent, such variations will remain stable because the same successful variations are more likely to be reinforced than less successful variations. However, behavioural variability can also be altered when subjected to certain controlling variables.

Avoidance Learning

Avoidance learning belongs to negative reinforcement schedules. The subject learns that a certain response will result in the termination or prevention of an aversive stimulus. There are two kinds of commonly used experimental settings: discriminated and free-operant avoidance learning.

Discriminated Avoidance Learning

In discriminated avoidance learning, a novel stimulus such as a light or a tone is followed by an aversive stimulus such as a shock (CS-US, similar to classical conditioning). During the first trials (called escape-trials) the animal usually experiences both the CS (Conditioned Stimulus) and the US (Unconditioned Stimulus), showing the operant response to terminate the aversive US. During later trials, the animal will learn to perform the response already during the presentation of the CS thus preventing the aversive US from occurring. Such trials are called "avoidance trials."

Free-operant Avoidance Learning

In this experimental session, no discrete stimulus is used to signal the occurrence of the aversive stimulus. Rather, the aversive stimulus (mostly shocks) are presented without explicit warning stimuli. There are two crucial time intervals determining the rate of avoidance learning. This first one is called the S-S-interval (shock-shock-interval). This is the amount of time which passes during successive presentations of the shock (unless the operant response is performed). The other one is called the R-S-interval (response-shock-interval) which specifies the length of the time interval following an operant response during which no shocks will be delivered. Note that each time the organism performs the operant response, the R-S-interval without shocks begins anew.

Two-process Theory of Avoidance

This theory was originally established to explain learning in discriminated avoidance learning. It assumes two processes to take place:

Classical conditioning of fear : During the first trials of the training, the organism experiences both CS and aversive US (escape-trials). The theory assumed that during those trials classical conditioning takes place by pairing the CS with the US. Because of the aversive nature of the US the CS is supposed to elicit a conditioned emotional reaction (CER) – fear. In classical conditioning, presenting a CS conditioned with an aversive US disrupts the organism's ongoing behaviour.

Reinforcement of the operant response by fear-reduction: Because during the first process, the CS signalling the aversive US has itself become aversive by eliciting fear in the organism, reducing this unpleasant emotional reaction serves to motivate the operant response. The organism learns to make the response during the US, thus terminating the aversive internal reaction elicited by the CS. An important aspect of this theory is that the term "avoidance" does not really describe what the organism is doing. It does not "avoid" the aversive US in the sense of anticipating it. Rather the organism escapes an aversive internal state, caused by the CS.

Verbal Behaviour

In 1957, Skinner published *Verbal Behaviour,* a theoretical extension of the work he had pioneered since 1938. This work extended the theory of operant conditioning to human behaviour previously assigned to the areas of language, linguistics and other areas. *Verbal Behaviour* is the logical extension of Skinner's ideas, in which he introduced new functional relationship categories such as intraverbals, autoclitics, mands, tacts and the controlling relationship of the audience. All of these relationships were based on operant conditioning and relied on no new mechanisms despite the introduction of new functional categories.

Four Term Contingency

Applied behaviour analysis, which is the name of the discipline directly descended from Skinner's work, holds that behaviour is explained in four terms: conditional stimulus (S^c), a discriminative stimulus (S^d), a response (R), and a reinforcing stimulus (S^{rein} or S^r for reinforcers, sometimes S^{ave} for aversive stimuli).

Operant Hoarding

Operant hoarding is a referring to the choice made by a rat, on a compound schedule called a multiple schedule, that maximises its rate of reinforcement in an operant conditioning context. More specifically, rats were shown to have allowed food pellets to accumulate in a food tray by continuing to press a lever on a continuous reinforcement schedule instead of retrieving those pellets. Retrieval of the pellets always instituted a one-minute period of extinction during which no additional food pellets were available but those that had been accumulated earlier could be consumed. This finding appears to contradict the usual finding that rats behave impulsively in situations in which there is a choice between a smaller food object right away and a larger food object after some delay.

An Alternative to the Law of Effect

However, an alternative perspective has been proposed by R. Allen and Beatrix Gardner. Under this idea, which they called "feed forward," animals learn during operant conditioning by simple pairing of stimuli, rather than by the consequences of their actions. Skinner asserted that a rat or pigeon would only manipulate a lever if rewarded for the action, a process he called "shaping" (reward for approaching then manipulating a lever). However, in order to prove the necessity of reward (reinforcement) in lever pressing, a control condition where food is delivered without regard to behaviour must also be conducted. Skinner never published this control group. Only much later was it found that rats and pigeons do indeed learn to manipulate a lever when food comes irrespective of behaviour. This phenomenon is known as autoshaping. Autoshaping demonstrates that consequence of action is not necessary in an operant conditioning chamber, and it contradicts the law of effect. Further experimentation has shown that rats naturally handle small objects, such as a lever, when food is present. Rats seem to insist on handling the lever when free food is available (contra-freeloading) and even when pressing the lever leads to less food (omission training). Whenever food is presented, rats handle the lever, regardless if lever pressing leads to more food. Therefore, handling a lever is a natural behaviour that rats do as preparatory feeding activity, and in turn, lever pressing cannot logically be used as evidence for reward or reinforcement to occur. In the absence of evidence for reinforcement during operant conditioning, learning which occurs during operant experiments is actually only Pavlovian (classical) conditioning. The dichotomy between Pavlovian and operant conditioning is therefore an inappropriate separation.

Sexual Imagery Viewing

A study by Platt, Khera and Deaner at Duke University (reported in *Current Biology* and online here), showed that male monkeys will give up privileges (in this case, juice, which is highly valued), to be allowed to see a female monkey's hindquarters.

Deaner and his team reported that monkeys would take a juice cut to look at powerful males' faces or the perineum of a female, but to persuade the monkeys to stare at subordinate males, the researchers had to bribe them with larger drinks.

The researchers stress that in monkey society, such behaviours have great social utility and we should therefore not simply reach the conclusion that "monkeys enjoy pornographic pictures". There is no evidence at this point that viewable pictures or movies of sexual activity are valued for their sexual enjoyment, although as noted above (Masturbation), there are reports that watching sex in real life may have such an effect. The subject of animals and sexual imagery is not yet well researched.

Problems with encouraging pandas to mate in captivity have been very common. However, showing young male pandas "panda pornography" is widely credited with a recent population boom among pandas in zoos.

Coercive Sex

Controversial interpretations and implications aside, sex in a forceful or apparently coercive context has also been documented in a variety of species. A notable example is bottlenose dolphins, where at times, a pod of bachelor males will 'corner' a female '...although what happens once the males have herded in a female, and whether she goes for one or all of them, is not yet known: the researchers have yet to witness a dolphin copulation.' The behaviour is also common in some arachnids (spiders), notably those whose females eat the males during sex if not tricked with food and/or tied down with threads, and in some herbivorous herd species or species where males and females are very different in size, where the male dominates sexually by sheer force and size.

Some species of birds appear to combine sexual intercourse with apparent violent assault; these include ducks, geese, and white-fronted bee-eaters. According to Emlen and Wrege (1986) forced copulations occur in this socially nesting species, and females must avoid the unwelcome attention of males as they emerge from their nest burrows or they are forced to the ground and mated with. Apparently, such attacks are made preferentially on females who are laying and who may thus mother their offspring as a result.

In 2007, research suggested that in the *Acilius* genus of water beetles (also known as "diving beetles"), an "evolutionary arms race" between the genders means that there is no courtship system for these beetles. "It's a system of rape. But the females don't take things quietly. They evolve counter-weapons." Cited mating behaviours include males suffocating females underwater till exhausted, and allowing only occasional access to the surface to breathe for up to six hours (to prevent them breeding with other males), and females which have a variety of body shapings (to prevent males from gaining a grip). Foreplay is "limited to the female desperately trying to dislodge the male by swimming frantically around."

Charles Siebert reports in his New York Times article *Elephant Crackup?* that:

> *Since the early 1990's, for example, young male elephants in Pilanesberg National Park and the Hluhluwe-Umfolozi Game Reserve in South Africa have been raping and killing rhinoceroses; this abnormal behaviour, according to a 2001 study in the journal Pachyderm, has been reported in "a number of reserves" in the region.*

Sex between Adults and Juveniles

It has also been recorded that certain species of mole will impregnate newborns of their own species. It is not clear if this is forceful or not. Similarly, the male stoat (Mustela erminea) will mate with infant females of their species. This apparently is a natural part of their reproductive biology – there is a delayed gestation period, so these females give birth the following year when they are fully grown.

A male spotted hyena which attempted to mate with a female which succeeded in driving it off, eventually turned to its ten-month-old cub, repeatedly mounting it and ejaculating on it. The cub sometimes ignored this and sometimes struggled 'slightly as if in play'. The mother did not intervene.

Infants and children in Bonobo societies are often involved in sexual behaviour.

Chapter 4

Sexual Cannibalism

Sexual cannibalism is a special case of cannibalism in which a female organism kills and consumes a male of the same species before, during, or after copulation. On rare occasions, these roles are reversed.

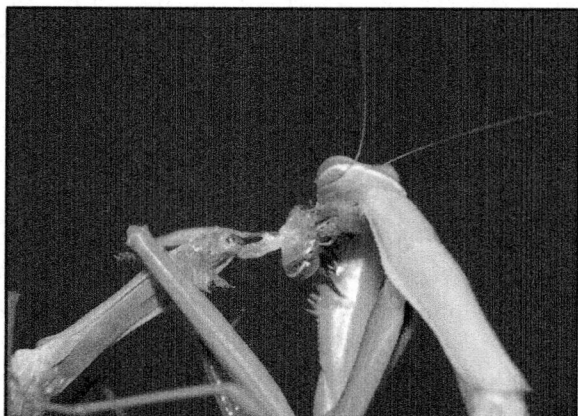

Figure 1: Sexual cannibalism in praying mantises: a female biting off the head of a male

Prevalence

Sexual cannibalism has been observed in arachnids, insects, and amphipods, with some evidence for its occurrence in gastropods, copepods, and cephalopods. Though it is rare overall, sexual cannibalism is common in most families of spiders and scorpions, and can affect population size and sex ratio. In most species in which it occurs, the female usually cannibalises the male, due to the large size of the female.

The significance of sexual cannibalism has been played down by some. Stephen Jay Gould argued that sexual cannibalism was too rare to be significant. However, males can be a significant food source for females, ultimately leading to increased fecundity.

Evolution and Maintenance

Figure 2: Female Chinese mantis eats a male copulating with her

Adaptive foraging hypothesis – Sexual cannibalism offers adult females direct, material benefits as part of an adaptive foraging strategy. An economic model showed that sexual cannibalism can be explained as the result of adaptive tradeoffs between the nutrient payoffs of cannibalising a male and the fertility benefits of mating with a male.

There are multiple predictions that can be derived from the adaptive foraging hypothesis. For a starving female, a courting male's value as food is higher than his value as a potential mate. As a result, hungry females are more likely to cannibalise than gorged females. Next, the female's previous mating status should affect the likelihood of precopulatory – but not post-copulatory – sexual cannibalism. Virgin females should be less likely to attempt to cannibalise males before sperm transfer than mated females. Next, the male's body mass determines his value as a meal, and thereby influences the chance of sexual cannibalism. Lastly, if precopulatory sexual cannibalism is an adaptive foraging method at the cost of mating, precopulatory attacks should be more successful than post-copulatory attacks, in a way that an attack is followed by a meal; otherwise, the most gainful approach for a female would be to copulate to secure sperm and then kill the male afterwards for food.

Aggressive spillover hypothesis – Sexual cannibalism may be the result of a "spillover" of aggression from juvenile life stages. High levels of aggression may be selectively advantageous in juveniles, giving them the ability to hunt and consume more prey and thereby grow more rapidly. If a high level of aggression is still present in adulthood, it can interfere with an individual's ability to mate and lead to sexual cannibalism. Positive behavioural correlations for aggression levels can develop across ontogenetic stages (juvenile-adult), as well as across behavioural contexts (foraging-mating). Natural selection then acts on the collection of correlated aggressive behaviours, instead of enhancing aggressive levels in each individual context.

The term "behavioural syndrome" describes such collections of correlated behaviours. Thus, this hypothesis represents an example of how a behavioural syndrome may produce behaviours that appear suboptimal when observed in an isolated context. It also predicts that females with an aggressive feeding history should grow larger and heavier than less aggressive females and are more likely to be cannibalistic. Some confirmation of the aggressive spillover hypothesis comes from a study of fishing spiders (*Dolomedes triton*).

Mate choice – Females may cannibalise unwanted males as a method of mate rejection or use cannibalism to regulate the timing of copulation. Direct mate choice via sexual cannibalism is a result of a specific discrimination by females, often by size, since size is strongly correlated with a variety of fitness characteristics. This is inferred from a female's diminished aggression towards certain males and increased aggression towards others.

For example, studies have shown that adult female wolf spiders, *Schizocosa uetzi*, prefer to mate with mates of a familiar phenotype. Males of an unfamiliar phenotype were more likely to be cannibalised. Researchers investigating direct mate choice suggest that: (a) females should react differently to some conspecifics (discriminate between them), (b) the discrimination should occur under natural conditions, (c) discrimination by the female between dissimilar males should cause differential reproductive success for males, (d) the female bias should be related to specific male morphological or behaviour traits, and (e) the variation among males in traits that are used by females to discriminate should be heritable.

Mistaken identity – The simplest explanation of sexual cannibalism is that it is a case of mistaken identity, in which a male is not acknowledged as a potential mate in time. However, in most spiders, elaborate courtship is thought to prevent this.

Polyandry advantage hypothesis – Polyandry is a mating system in which one female mates with more than one male. The polyandry advantage hypothesis predicts that females will attempt to prevent male monopolisation of their eggs through sexual cannibalism.

Environmental and evolutionary circumstances – Spider species vary extensively in the degree of sexual size dimorphism, which is a result of variation among species in the strengths of fecundity selection acting to increase female size and viability selection, along with other selective pressures acting to maintain or decrease male size. As a result, this degree of sexual size dimorphism has consequences for sexual cannibalism, since it will affect male vulnerability to female attacks.

Male Self-sacrifice

In some cases, sexual cannibalism may characterise an extreme form of male monogamy, in which the male will sacrifice itself to the female. Males may gain reproductive success from being cannibalised by either providing nutrients to the female (indirectly to the offspring), or through enhancing the probability that their sperm will be used to fertilize the female's eggs. Although sexual cannibalism is fairly common in spiders, male self-sacrifice has only been reported in six genera of araneoid spiders. In the most studied example, the Australian redback widow spider (*Latrodectus hasselti*, Theridiidae), males position their abdomen in front of the female's chelicerae in a somersault behaviour and are consumed during copulation; sexual cannibalism occurs in about 65% of matings.

Relating to the adaptive self-sacrifice theory, cannibalised redback males double their paternity compared to rival males. In addition, male orb-weaving spiders of the species *Argiope aemula* and *Argiope aurantia*, which are sexually cannibalistic, die immediately after mating, even if they are not attacked. Since they appear to expect no future mating success, they may have nothing to lose by being cannibalised, making the benefits of sexual cannibalism outweigh the costs.

Reversed Sexual Cannibalism

Reversed sexual cannibalism may transpire primarily in species in which males are larger than females. King cobras do feed on other snakes apart from rodents and other natural prey of snakes. Male king cobras are larger in size than their female counterparts and may consume the female or mate with them. Male amphipods of the species *Gammarus pulex* and *G. duebeni celticus* may eat smaller females, especially at the moult, when they are most vulnerable. Likewise, female paddle crabs, *Ovalipes catharus*, are also defenceless at the moult when their shell is soft, and may be cannibalised during or after mating. In *Gammarus* species, sexual cannibalism would appear to be naturally opportunistic and not have any mate choice consequences.

Evidence has also been found of a male eating a female in the mantid *Tarachode afzeli*. Males of the water spider *Argyronetia aquatica* are larger than females, prefer to mate with larger females, and may cannibalise females that are smaller than themselves. Females prefer to mate with larger males, but are more cautious of them; they flee from larger males more often than from males that are closer to their own size.

The vulnerability to sexual cannibalism is a key in *A. aquatica* and would propose that an indirect mechanism, instead of direct mate choice, yields the size-dependent reversed sexual cannibalism that is observed. While the presence of the male-biased size dimorphism and the relative size of the female determining her vulnerability, smaller females are likely to be cannibalised. In all cases of reversed sexual cannibalism, there appears to be little hope that the female will produce viable offspring for the male, or the need for food exceeds the fitness gain that can be achieved through mating. Since most males will tend towards promiscuity in order to maximise their reproductive output, it is difficult to visualise mate choice through sexual cannibalism for males, unless the sex ratio is strongly female-biased, offering the males the opportunity to be choosy or the sex roles are reversed.

Traumatic Insemination

Traumatic insemination, also known as hypodermic insemination, is the mating practice in some species of invertebrates in which the male pierces the female's abdomen with his penis and injects his sperm through the wound into her abdominal cavity (hemocoel). The sperm diffuse through the female's hemolymph, reaching the ovaries and resulting in fertilization. The process is detrimental to the female's health. It creates an open wound which impairs the female until it heals, and is susceptible to infection. The injection of sperm and ejaculatory fluids into the hemocoel can also trigger an immune reaction in the female. Bed bugs, which reproduce solely by traumatic insemination, have evolved a pair of sperm-receptacles, known as the spermalege. The spermalege reduce the damage to the female bed bug during traumatic insemination.

The evolutionary origins of traumatic insemination are disputed. Although it evolved independently in many invertebrate species, traumatic insemination is most highly adapted and thoroughly studied in bed bugs, particularly *Cimex lectularius*. Traumatic insemination is not limited to male-female couplings, or even couplings of the same species. Both homosexual and inter-species traumatic inseminations have been observed.

Some species of insect have evolved spiny penises, which damage the female reproductive tract. This has led to females using various techniques to resist being bred.

In humans and other complex life forms, blood and lymph circulate in two different systems, the circulatory system and lymphatic system, which are enclosed by systems of capillaries, veins, arteries, and nodes. This is known as a closed circulatory system. Insects, however, have an open circulatory system in which blood and lymph circulate unenclosed, and mix to form a substance called hemolymph. All organs of the insect are bathed in hemolymph, which provides oxygen and nutrients to all of the insect's organs.

Following traumatic insemination, sperm can migrate through the hemolymph to the female's ovaries, resulting in fertilization. The exact mechanics vary from taxon to taxon. In some orders of insects, the male genitalia (paramere) enters the female's genital tract, and a spine at its tip pierces the wall of the female's bursa copularix. In others, the male penetrates the outer body wall. In either case, following penetration, the male ejaculates into the female. The sperm and ejaculatory fluids diffuse through the female's hemolymph. The insemination is successful if the sperm reach the ovaries and fertilize an ovum.

Female resistance to traumatic insemination varies from one species to another. Females from some genera, including *Cimex*, are passive prior to and during traumatic insemination. Females in other genera resist mating and attempt to escape. This resistance may not be an aversion to pain caused by the insemination, as observational evidence suggests that insects do not feel pain.

Research into the paternity of offspring produced by traumatic insemination has found "significant" last-sperm precedence. That is, the last male to traumatically inseminate a female tends to sire most of the offspring from that female.

Evolutionary Adaptation

Many reasons for the evolutionary adaptation of traumatic insemination as a mating strategy have been suggested. One is that traumatic insemination is an adaptation to the development of the mating plug, a reproductive mechanism used by many species. Once a male finishes copulating, he injects a glutinous secretion into the female's reproductive tract, thereby "literally glu[ing] her genital tract closed". Traumatic insemination allows subsequent males to bypass the female's plugged genital tract, and inject sperm directly into her circulatory system.

Others have argued that the practice of traumatic insemination may have been an adaptation for males to circumvent female resistance to mating to eliminate courtship time, allowing one male to inseminate many mates when contact between them is brief; or that it evolved as a new development in the sperm competition as a means to deposit sperm as close to the ovaries as possible.

This bizarre method of insemination probably evolved as male bed bugs competed with each other to place their sperm closer and closer to the mother lode of eggs, the ovaries. Some male insects evolved long penises with which they enter the vagina but bypass the female's storage pouch and deposit their sperm further upstream close to the ovaries. A few males, notably among bed bugs, evolved traumatic insemination instead, and eventually this strange procedure became the norm among these insects.

It has recently been discovered that members of the plant bug genus *Coridromius* (Miridae) also practice traumatic insemination. In these bugs, the male intromittent organ is formed by the coupling of the aedeagus with the left paramere, as in bed bugs. Females also exhibit paragenital modifications at the site of intromission, which include grooves and invaginated copulatory tubes to guide the male paramere. The evolution of traumatic insemination in *Coridromius* represents a third independent emergence of this form of mating within the true bugs.

Health Repercussions

While advantageous to the reproductive success of the individual male, traumatic insemination imposes a cost on females: reduced lifespan and decreased reproductive output. "These [costs] include (i) repair of the wound, (ii) leakage of blood, (iii) increased risk of infection through the puncture wound, and (iv) immune defence against sperm or accessory gland fluids that are introduced directly into the blood."

The male bed bug penis has been shown to carry five (human) pathogenic microbes, and the exoskeleton of female bed bugs nine, including *Penicillium chrysogenum, Staphylococcus saprophyticus*,

Stenotrophomonas maltophilia, Bacillus licheniformis, and *Micrococcus luteus.* Tests with blood agar have shown some of these species can survive *in vivo.* This suggests infections from these species may contribute to the increased mortality rate in bed bugs due to traumatic insemination.

The successive woundings each require energy to heal, leaving less energy available for other activities. Also, the wounds provide a possible point of infection which can reduce the female's lifespan. Once in the hemolymph, the sperm and ejaculatory fluids may act as antigens, triggering an immune reaction.

There is a tendency for dense colonies of bed bugs kept in laboratories to go extinct, starting with adult females. In such an environment, where mating occurs frequently, this high rate of adult female mortality suggests traumatic insemination is very detrimental to the female's health. The damage done, and the (unnecessarily) high mating rate of captive bed bugs, have been shown to cause a 25% higher-than-necessary mortality rate for females.

Bed Bug Adaptation

The effects of traumatic insemination are deleterious to the female. Female bed bugs have evolved a pair of specialised reproductive organs ("paragenitalia") at the site of penetration. Known as the ectospermalege and mesospermalege (referred to collectively as spermalege), these organs serve as sperm-receptacles from which sperm can migrate to the ovaries. All bed bug reproduction occurs via traumatic insemination and the spermalege. The genital tract, though functional, is used only for laying fertilized eggs.

The ectospermalege is a swelling in the abdomen, often folded, filled with hemocytes. The ectospermalege is visible externally in most bed bug species, giving the male a target through which to impale the female with the paramere. In species without an externally visible ectospermalege, traumatic insemination takes place over a wide range of the body surface.

Exactly why males 'comply' with this aspect of female control over the site of mating is unclear, especially as male *P. cavernis* appear to be able to penetrate the abdomen at a number of points independent of the presence of an ectospermalege. One possibility is that mating outside the ectospermalege reduces female fecundity to such an extent that the mating male's paternity is significantly reduced ... The ectospermalege appears to act as a mating guide, directing the male's copulatory interest, and therefore damage, to a restricted area of the female's abdomen.

The mesospermalege is a sac attached to the inner abdomen, under the ectospermalege. Sperm is injected through the male's penis into the mesospermalege. In some species, the ectospermalege directly connects to the ovaries—thus, sperm and ejaculate never enters the hemolymph and thus never trigger an immune reaction. (The exact characteristics of the spermalege vary widely across different species of bed bugs.) The spermalege are generally found only in females. However, males in the *Afrocimex* genus possess an ectospermalege. Sperm remains in the spermalege for approximately four hours; after two days, none remains.

Male bed bugs have evolved chemoreceptors on their penises. After impaling a female, the male can "taste" if a female has been recently mated. If he does, he will not copulate as long and will ejaculate less fluid into the female.

Use in the Animal Kingdom

Although traumatic insemination is most widely practiced among heteropterans (typical bugs), the phenomenon has been observed across a wide variety of other invertebrate taxa. These include:

- *Oxyurida* (nematodes) – Traumatic insemination has been observed in pinworm genera including *Auchenacantha*, *Citellina*, *Passalurus*, and "probably" *Austroxyris*

- *Acanthocephala* (parasitic, thorny-headed worms) – The presence of mating plugs on the sides of *Pomphorhynchus bulbocolli* suggests traumatic insemination occurs in this species. Because these parasites cannot move after anchoring themselves to a host's intestine, traumatic insemination may have evolved to compensate for their immobility.

- *Rotifera* (wheel animalcules) – In the *Brachionus* genus, the male pierces the syncytial integument (equivalent to skin) and injects sperm; in *Asplanchna brightwelli* the male secretes an enzyme which breaks down the female integument and injects sperm through the hole.

- *Turbellaria* (free living flatworms) – Hermaphroditic flatworms reproduce by "penis fencing". Individuals "fence" with penises, attempting to use their penis to pierce the skin of the other and inject sperm. The 'loser' is the flatworm which is inseminated and must bear the energy costs of reproduction. One study of *Pseudoceros bifurcus* found "Most inseminations were unilateral. Even when reciprocal penis insertion could be achieved by the second partner, the first to inseminate obtained a longer injection time than the second."

- Gastropod snails
- *Strepsiptera* (twisted-winged parasites) – In *Xenos vesparum*, fertilization can occur either via extragenital ducts, or by traumatic insemination into the hemocoel.
- *Drosophila* (flies) – Ejaculates are injected through the body wall into the genital tract, not the abdomen.
- *Opisthobranchia* (sea slugs) – Characterised by "repeated small injections into the dorsal surface of the partner, interrupted by synchronised circling movements", culminating in a standard genital insemination.
- *Harpactea* (spiders) – The male of the spider species *Harpactea sadistica* pierces the female's body cavity and inseminates her ovaries directly.

Homosexual Traumatic Insemination

Traumatic insemination is not limited to male–female couplings. Male homosexual traumatic inseminations have been observed in *Xylocoris maculipennis* and the *Afrocimex* genus.

In the genus *Afrocimex*, both species have well developed ectospermalege (but only females have a mesospermalege). The male ectospermalege is slightly different from that found in females, and amazingly enough, Carayon (1966) found that male *Afrocimex* bugs suffer actual homosexual traumatic inseminations. He found the male ectospermalege often showed characteristic mating scars, and histological studies showed "foreign" sperm were widely dispersed in the bodies of these homosexually mated males. Sperm cells of other males were, however, never found in or near the male reproductive tract. It therefore seems unlikely that sperm from other males could be inseminated when a male that has himself suffered traumatic insemination mates with a females. The costs and benefits, if any, of homosexual traumatic insemination in *Afrocimex* remain unknown.

Klaus Reinhardt of the University of Sheffield and colleagues observed two morphologically different kinds of spermalege in *Afrocimex constrictus*, a species in which both male and females are traumatically inseminated. They found females use sexual mimicry as a way to avoid traumatic insemination. In particular, they observed males, and females who had male spermalege structures, were inseminated less often than females with female spermalege structures.

In *Xylocoris maculipennis*, a flower bug, after a male traumatically inseminates another male, the injected sperm migrate to the testes. (The seminal fluid and most of the sperm are digested, giving the inseminated male a nutrient-rich meal.) It has been suggested, although there is no evidence, that when the inseminated male ejaculates into a female, the female receives both males' sperm.

Interspecies Traumatic Insemination

Cases of traumatic insemination between animals of different species will sometimes provoke a possibly lethal immune reaction. A female *Cimex lectularius* traumatically inseminated by a male *C. hemipterus* will swell up at the site of insemination as the immune system responds to male ejaculates. In the process, the female's lifespan is reduced. In some cases, this immune reaction can be so massive as to be almost immediately fatal. A female *Hesperocimex sonorensis* will swell up, blacken, and die within 24–48 hours after being traumatically inseminated by a male *H. cochimiensis*.

Similar Mating Practices

In the animal kingdom, traumatic insemination is not unique as a form of coercive sex. Research suggests, in the *Acilius* genus of water beetles, there is no courtship system between males and females. "It's a system of rape. But the females don't take things quietly. They evolve counter-weapons." Cited mating behaviours include males suffocating females underwater till exhausted, and allowing only occasional access to the surface to breathe for up to six hours (to prevent them breeding with other males), and females which have a variety of body shapes (to prevent males from gaining a grip). Foreplay is "limited to the female desperately trying to dislodge the male by swimming frantically around".

"Rape behaviour" has been observed in a number of duck species. In the blue-winged teal, "rape attempts by paired males may occur at any time during the breeding season." Cited reasons for this being beneficial to the paired males include successful reproduction, and chasing away intruders from their territory. Bachelor herds of bottlenose dolphins will sometimes gang up on a female and coerce her to have sex with them, by swimming near her, chasing her if she attempts to escape, and making vocalised or physical threats. In the insect world, male water striders unable to penetrate her genital shield, will draw predators to a female until she copulates.

Interlocus Sexual Conflict

Interlocus sexual conflict is the idea of conflict occurring because the action of one sex decreases the fitness of the other sex as a by-product. For example in fruit flies, males have evolved particular seminal compounds that function in the context of sperm competition, but at the same time reduce female fitness even though that is not its primary function in response. One sex may have developed a counter-adaptation to the action of the other sex. For example, female bed bugs evolved paragenital structures that function to reduce the damages inflicted by males' needle-like intromittent organs.

The conflict between sexes is the result of the sex-specific relationship between mating rate and fitness. The positive relationship between mating rate and fitness is much stronger for males than for females, because of anisogamy. However, conflict will arise if the fitness of one sex is decreased by the action of the other sex.

Mating rate is influenced by the interactions between loci that are expressed for each sex. For instance, locus A is expressed in males, while locus B is expressed in females, the outcome of this interaction can vary. The alleles of locus A are favoured and will increase the rate of mating for male carriers, resulting in the spread of these alleles in the population. For instance, in fruit flies, males have toxic sperm representing locus A. Locus A of the male fruit flies reduces female fitness. These particular alleles along with their phenotypes results in the mating rate increasing for males. Counter adaptive traits in females probably evolved to reduce the harm of male detrimental traits, and may be the result of a mutation. Mutations occur randomly, and if this causes a beneficial effect such as increasing female fitness by countering male actions, then these traits will spread in the population. Ultimately, this event can cause the alleles of locus B to affect the selection for alleles of locus A.

Sexual conflict can be explained by one of the hypothesis known as the chase-away sexual selection model. This concept involves the process of males having traits that helps to tap into a pre-existing sensory bias of females. This affects the female preference for a particular mate (female choice). Males who manipulate females into mating with them, may not offer genetic benefits compared to other males of his species, resulting in females evolving resistance towards males. Selection would favour males that are capable of overcoming female resistance due to his exaggerated traits. This may result in a cyclic process.

Interlocus sexual conflict can be the result of what is called a perpetual cycle. The perpetual cycle begins with the traits that favour male reproductive competition, which eventually manifests into male persistence. The traits of the favour males will cause a reduction in the fitness of females due to their persistence. Following this event, females may develop a counter-adaptation. In other words, females may develop favourable traits that reduce the direct costs implemented by males, which is known as female resistance. After this event, there is a decline in the fitness depression in females, and the cycle returns to the beginning with favourable traits in males. Interlocus sexual conflict reflects interactions among mates that can be beneficial or not and can be explained through evolutionary concepts.

Sensory exploitation by males is one of the mechanisms that involve males trying to overcome female reluctance. Chase-away selection is the outcome of this, resulting in a co-evolutionary arms race. Other mechanisms include traumatic insemination, forced copulation, sexual cannibalism, penis fencing, love darts and others.

Evolutionary Theories

Interlocus sexual conflict involves numerous evolutionary concepts that are applied to a wide range of species in order to provide explanations for the occurrence of interactions among sexes. The conflict between the interactions of male and females can be described as an ongoing evolutionary arms race.

According to Darwin (1859), sexual selection is when individuals are favourable over others of the same sex in the context of reproduction. Sexual selection and sexual conflict are connected because males usually mate with some or many females compared to females that do not mate with many males.

It is hypothesized that sexual conflict may be the result of sensory exploitation implemented by males. Co-evolutionary process describes chase-away selection as a result of sensory exploitation by males. Female choice is a possible reason why males may be able to tap into or exploit females' sensory biases. For example, females may behave in ways that are considerably biased towards mating and fertilization success due the attractiveness of males who exhibit a deceptive or exaggerated trait. Since some male traits are detrimental to females, it may result in the female leaning towards the preference for male resistant traits or becoming insensitive to these traits. Sexually antagonistic co-evolution entails the cyclic process between the exaggerated (persistent) traits and the resistant traits between the sexes. The exaggerated traits of males can influence the evolutionary responses in females. If the outcome consists of the spread of male traits that decreases the fitness in females, this will result in an evolutionary alteration in the preference of females. Therefore, sexual conflict between sexes can influence the preference females may have towards males.

Female's Resistance

Female resistance is an evolutionary concept that involves females losing attractiveness towards males. Sexual conflict could possibly be the result of females' loss of attractiveness. This concept can be supported by the water strider and pygmy fish.

Male water striders exhibit forced copulation (or rape) on the female. As a result the female will struggle with the male in order to reduce the detrimental effects. Female's struggling is a by-product of female resistance.

The population of pygmy fish *X. pygmaeus* or pygmy sword-tail fish at the time consists of small males. A study decided to test female choice, they decided to use large hetero-specific males. What they found was that female pigmy swordtail fish favoured large size males. This revealed how females change their preference from small males to large males. Females may be resistant to small males because of their loss of attractiveness. It was observed that this pattern for female preference for large male body size actually disappeared among populations that consisted of smaller males. It was concluded this behaviour is caused by female resistance and not her preference for larger body size males.

Sperm Competition

Sperm competition is an evolutionary concept developed by Geoff Parker (1970). Sperm competition describes that ejaculates belonging to different males will compete to fertilize the egg of the female. Sperm competition selects for offensive and defensive traits. Offensive sperm competition consists of males displacing sperm from the previous male. Conversely, defensive sperm competition consists of males preventing females from remating by delaying them, or restricting their interests in other males. Sperm competition can be exhibited throughout behavioural, morphological and physiological male adaptations.

Some examples of behavioural adaptations are mate guarding or forced copulation. Morphological adaptations may include male claspers, altered genitalia (i.e. spiky genitals) and copulatory plugs (mating plugs). Finally, physiological adaptations may consist of toxic sperm, or other chemicals within the seminal fluid that causes a delay in female's ability to remate.

Sexual conflict can be exhibited through males targeting other males in the context of sperm completion. For example, Iberian Rock Lizard (*Lacerta monticola*) males have been known to create mating plugs that are described to be hard. These mating plugs are placed within the female cloaca instantly after copulation.

The hypothesis was that mating plugs can lower the ability of a female to be attracted to other males and that the function of mating plugs may serve as a "chastity belt." The study revealed, there was no evidence to support the hypothesis, and that males were able to displace the mating plugs from other males. There is no direct conflict between males and females, but males may evolve manipulative traits because rival males are removing mating plugs.

Males will adapt or develop different behaviours for paternity assurance. A study of sperm competition revealed that there was a positive relationship between testis size and levels of sperm competition within groups. It was also evident that the larger the accessory reproductive glands, the seminal vesicles, and anterior prostate were, had a higher level of sperm competition among males. The larger the mating plugs were produced were the less likely they were removed.

Advantages

Males: Males inflicting harm to females is seen as a by-product of male adaptation in the context of sperm competition. The advantages to males may include: a) decreasing the likelihood of females remating, b) the ability of providing a larger investment in terms of offspring, c) sperm maintenance, and d) sperm storage.

These advantages are seen throughout all variations of mate traits such as toxic sperm, spiky genitalia, forced copulation, sexual cannibalism, penis fencing, love darts, mate guarding, harassment/ aggressive behaviour, and traumatic insemination.

Hermaphrodites

Hermaphrodites are described as organisms that have the ability to interchangeably shift from males to females or vice versa. When compared to separate sexes or in this case gonochorists, hermaphrodites benefits much more in terms of reproduction. For example, marine flatworms exhibit penis fencing. Penis fencing involves two mates fighting in order to stab the other mate first. The mate that is stabbed first will take on the female role and produce the offspring.

Hermaphrodites can behave as males which are the sperm donors or females which are usually the recipients of sperm, or both. There are cases where hermaphrodites have the potential to fertilize their own eggs, but this is usually rare. Most hermaphrodites take on the role as male or female in order to effectively reproduce. Sexual conflict over mating can cause hermaphrodites to either cooperate or display aggressive behaviour in the context of gender choice.

Hermaphrodites are able to increase their offspring's fitness due to outbreeding. Another reason is that hermaphrodites digest sperm, which provides nutritional benefits. The ejaculates are beneficial to the female because the ejaculates contain a variety of substances that can signal the activation of reproduction.

Disadvantages

Females: Females can experience a wide range of detrimental effects from males. This may include: a) longevity reduction, b) distortion in feeding behaviours, that can at times increase the food intake as seen in *Drosophila* fruit flies, c)increased risk of infection, d) wound repair consuming energy, e) males manipulating female's reproductive schedules, f) being susceptible to predators, and g) females' immune response being reduced.

Sexual Conflict Examples

Mate's tactics can consist of a wide range of behaviours from morphologically to behaviourally or even physiologically. Traumatic insemination can be divided into intra-genitalic and extra-genitalic traumatic insemination. Intra-genitalic is described as cryptic since it involves males placing their intromittent organ into the genitalia of the female. Following that event, the male will pierce the inner wall of the female's genitalia in order to insert his sperm. Extra-genitalia can be seen as transparent due to males' method of piercing the body wall of the female in a region that is distant from the female's genitalia.

Intra-genitalic traumatic insemination is seen in Drosophila fruit flies, while extra-genitalic traumatic insemination is in species such as bed bugs, spiders, beetles and hermaphrodites.

Traumatic Insemination

Figure 3: A male bed bug (*Cimex lectularius*) traumatically inseminates a female bed bug (top). The female's ventral carapace is visibly cracked around the point of insemination.

Traumatic insemination describes the mate's tactics of piercing a female and depositing sperm in order to ensure the paternity success. Traumatic insemination in this sense incorporates species who displays extra-genitalic traumatic insemination. Males have a needle like intromittent organ. Examples includes bed bugs, bat bugs and spiders.

Bed bugs *Cimex lectularius*, for example, males initiate mating by climbing onto the female and piercing her abdomen, the male will then directly inject his sperm along with the accessory gland fluids into the female's blood. As a result the female will have a distinct melanised scar in the region where the male pierced. It was observed that mounting or piercing by males can result in males not only piercing females but other males and nymphs. The females may suffer detrimental effects which can include blood leaking, wounds, the risk of infection, and the immune system having difficulty in fighting off sperm located in the blood.

A study focused on the mating effects of bed bugs of other species such as female *Hesperocimex sonorensis* and a male *Hesperocimex cochimiensis*. It was observed that *H. sonorensis* females died in a period of 24 to 48 hours after mating with *H. cochimiensis* males. When examining the females, it was evident that their abdomens were blackened and swollen due to enormous amounts of immunoreactions. There is a direct relationship between the increase of mating and the decrease in female's lifespan. It can be argued that female bed bugs that suffer traumatic insemination are not dying due to a male piercing them, but to a microbe.

Female bed bug mortality rate due to traumatic insemination could be more STD related than just the open wound. The same environmental microbes were found on the male's genital was also found within the female. A study found a total of nine microbes, with five microbes actually causing mortality of females during copulation.

African bat bugs *Afrocimex constrictus* also perform extra-genitalic traumatic insemination. Males will puncture the female outside her genital and ultimately inseminate them. It was observed that both males and females suffer from traumatic insemination. Males suffer from traumatic insemination because they expressed female like genitals, and were often at times mistaken for females. Females also displayed what is called polymorphism because some females had distinct "female-like" genitals while others had a "male-like" appearance. The results showed that males along with females who had "male-like" genitals suffer less traumatic insemination compared to the distinct females. Female's polymorphism could in fact be a result of evolution due to sexual conflict.

Male spiders *Harpactea sadistica* perform extra-genitalic traumatic insemination with his needle-like intromittent organ that will puncture the female's wall, resulting in direct insemination. Males also puncture females with their cheliceral fangs during courtship. Females have an atrophied spermathecae (sperm-storage organ). The sperm storage organ removes sperms from males who mate later, which reflects cryptic female choice. Cryptic female choice entails a female's opportunity to choose which sperm to fertilize her eggs with. It has been suggested that males may have developed this aggressive mate tactic as a result of female's sperm storage organ.

Toxic Ejaculation

Toxic ejaculation is most associated with *Drosophila melanogaster* fruit flies. *Drosophila* fruit flies exhibit toxic ejaculation along with intra-genitalic traumatic insemination. The males will place his intromittent organ within the female genitalia, following the piercing of her inner wall to inject toxic sperm. *Drosophila* males benefit highly from toxic ejaculation because the female will only produce his offspring.

The substance in the toxic ejaculate that has been linked to detrimental effects to females is known as accessory gland proteins (Acps). Acps are found in males' seminal fluid. After Acps are transferred into the female, this causes a change in her behaviour and physiology. A study found that females who received Acps products from males suffered a decrease in their lifespan.

Currently it has been estimated that there are more than 20 different genes that code for Acps. It has been approximated that male fruit flies contain a range of 80 to 100 different Acps. Acps genes have been found in a variety of species and genera. Acps has been described as displaying a conservation function because it reserves protein biochemical classes within the seminal fluid. Since Acps exist in a variety of species, fruit flies may have evolved a higher concentration for mating success.

Drosophila hibisci species' mating tactics are mating plugs rather than traumatic insemination. The mating plugs of *Drosophila hibisci* are characterised as a gelatinous, hard composite that adheres to the uterus of the female in the event of copulation. A study tested two hypotheses concerning mating plugs: a) if they were nutritional gifts for females to digest that will provide maintenance of the eggs during maturation, or b) if they could serve as a chastity device to prevent sperm of rivals. What the study found was that mating plugs had no effect on female nutrition and serve as an enforcement device against rival males. Although this species of fruit flies (*Drosophila hibisci*) found success in mating plugs, they are ineffective for other species *Drosophila*. A study found that males who insert their mating plugs within females were unable to prevent females from remating just four hours after mating. Therefore, the assumption can be made that since mating plugs are insufficient, that male *Drosophila melanogaster* develop other male adaptations to compensate for this defect such as intra-genitalic traumatic insemination in order to directly deposit their sperm.

Spiky Genitals

The Bruchid beetle or bean weevil *Callosobruchus maculatus* males are known to express extra-genitalic traumatic insemination on females. The male Bruchid beetle's intromittent organ is described as having spines that are used to pierce the reproductive tract of the female.

Males who had multiple copulations with the same female resulted in a greater damage to her genital. However, those same males transferred a small quantity of ejaculates compared to the virgin males. It was also observed that males who participated in copulation with females, at times deposit no sperm through the wounds they created on the females.

The study also found that females who mated with more than one male would suffer a higher mortality. It was also noted that females had a decrease in longevity as a result of receiving a large single ejaculate from males.

However, females who received a total of two ejaculates were less likely to die compare to those that received just one ejaculate. The assumption could be made that females who mated 48 hours after the first copulation were lacking nutrition since as adults they do not drink or eat. The ejaculate that was provided after the second copulation was nutritionally beneficial and lengthen female's longevity. By increasing longevity of females, females were able to produce more offspring.

Females who mated with virgin males were less likely to obtain genital damage compared to regular males. It was suggested that factors contributing to male virgins being less harmful were ejaculate size and the amount of sperm contained.

Penis Fencing

Hermaphrodites such as marine flatworms may display an extra-genitalic traumatic insemination known as penis fencing. The idea of penis fencing consists of both mates competing with one another in order to puncture the other mate first. The mate that gets punctured first will be selected to reproduce. In penis fencing, males described as having "stylet-shaped penises" will aggressively try to stab the other mate. Once the mate successful punctures the other mate, sperm is then inseminated. The idea is that potential males are forcing female gender roles on the other mate.

Forcing gender roles does not only occur in flatworms, but in other hermaphrodites such as the banana slug *Ariolimax spp.* In banana slugs, copulation occurs following sperm deposit in which gender between the slugs is selected. The banana slug, who displays a female role has a reproductive tract that has the ability to grip the male's penis. As a result, both genders are locked together during copulation. During separation, if unable to disconnect, after repeated attempts, aprophallation may occur. Aprophallation describes an incidence where the penis may be trapped, resulting in the penis of the male being bitten or even consumed by the female. The mate who lost their penis is unable to perform as the male role during mating. Compare to penis fencing, banana slugs are prone to force gender roles on other mates just as well.

Love Darts

Hermaphroditic gastropod snails' mate tactic consists of love darts. The love darts are described as a sharp "stiletto," that is created by the males. The love darts are shot at the females during courtship. A single love dart is shot at a time, due to the lengthy process of regenerating.

The model organism that is studied in terms of love darts in snails of the genus *Helix*. It was observed that snails that rub against their mates, will forcefully place the love dart into their mate. It have been shown that darts may aid in mating, but not necessary ensure mating success.

Love darts do in fact aid in mating success. Hermaphroditic snails will selectively take on a female or male role. Snails transmitted darts into these females so that they would store much more sperm (about twice as much) compared to males who weren't as successful. Males who successfully hit females with love darts had a higher paternity assurance. Many snails who are inflicted by love darts suffer open wounds and sometimes death.

Forced Copulation

Forced copulation occurs in a wide range of male species and may elicit behaviours such as aggression, harassment and grasping. In the time prior or during copulation, females suffer detrimental effects due to male's forceful mate tactics. Ultimately, females are being forced to copulate without their permission (i.e. rape).

Harassment

Harassment is a behaviour that is displayed in the event of or prior to forced copulation. Males may display a stalker like behaviour, in which he is following the female at a distance in preparation to attack. In the Malabar ricefish *Horaichthys setnai* (Beloniformes) males would harass females of interest from a distance. This stalker like behaviour may consist of swimming below or behind the females, and even following them at a distance. When the male Malabar ricefish is ready to copulate, he will dash at high speed towards the female and release his club-shaped organ, gonopodium also known as an anal fin. The purpose of the gonopodium is to deliver the spermatophore. The male will take his gonopodium and forcefully place it near the female genitalia. The sharp end of the spermatophore will stab the female's skin. As a result, the male is firmly attached to the female. Following this event, the male's spermatophore will burst, leading to the release of the sperm that will travel towards the female's genital opening.

Grasping

Forced copulation can lead to aggressive behaviours such as grasping. Males will express grasping behaviours in the event of copulation towards a desired female. Darwin (1871) described males with grasping qualities as "organs for prehension." His view was that males perform these aggressive behaviours in order to prevent the female from leaving or escaping. The purpose of males' grasping devices is to increase the duration of copulation along with restricting females from other males. Grasping traits can also be considered as a way of males expressing mate-guarding. Examples of species with grasping traits are water striders and diving beetles.

During forced copulation, male water striders genus *Gerris* will attack females. As a result, a struggle will occur because the female is resistant. When the male water strider is successfully attached to the female, the female will carry the male during and after copulation. This can be energetically costly to the female because she has the heavy weight of the male at the same time she is gliding on the water surface. The speed of the female is usually reduced by 20% when the male is attached. The purpose of long copulation is for the male to achieve paternity assurance in order to restrict the female from other males. Long periods of copulation can strongly affect females because females will depart from the water surface after mating and discontinue foraging. The duration of copulation can be extremely long. For water strider *Aquarius najas* it was a total of 3 months. For water strider *Gerris lateralis* the time ranged from 4 to 7 minutes.

In water strider *Gerris odontogaster*, males have an abdominal clasping mechanism that grasps females in highly complex struggles before mating. Males that have clasps that are longer than those of other males were able to endure more somersaults by resistant females and achieved mating success. Males' genital structures had a particular shape to aid in female resistance.

Water striders *G. gracilicornis* have a behavioural mechanism and grasping structures. Male water striders use what is called an "intimidating courtship". This mechanism involves males using a signal vibration to attract predators in order to manipulate females to mate. Females face more risks of being captured by predators since they do lots of idling on the water surface. If a male was attached to the female, it is less likely for the male to be harm by the predators because he is rested on top of the female. Therefore, males will do leg tapping in order to create ripples in the water to attract predators. The female will become fearful causing her to be less resistant towards the male. As a result, copulation occurs faster, in which the male stops signalling.

Male water striders *Gerris odontogaster* have grasping structures that can prolong copulation depending on the size of their abdominal processes. Males who had longer abdominal processes were able to restrain females compared to males who had shorter abdominal processes. It was evident that females developed a preference for males with longer abdominal processes which expresses her bias towards mating. This is female choice because females prefer males who are stronger, in order to have offspring with those genes.

In diving beetles *Dytiscidae*, males will approach females in the water with a grasping mechanism before copulation. When this occurs, females would repeatedly resist. Males evolved an anatomical advantage towards grasping. Apparently males have a particular structure located on their tarsae that enhances grasping of female's anatomical structures, pronotum and elytra, which are located on her dorsal surface.

Sexual Cannibalism

Sexual cannibalism contradicts the traditional male-female relationship in terms of interlocus sexual conflict. Species that exhibit in sexual cannibalism are spiders and invertebrates. Sexual cannibalism involves females slaying and consuming males during courtship or copulation. Examples are the gift-giving spider and the funnel-web spider.

A possible explanation for sexual cannibalism occurring across taxa is "paternal investment". This means that females would kill and consume males after sperm exchange in order to enhance the quality and amount of her offspring. Males being consumed by females would serve as a blood meal since they are volunteering their soma. During sexual cannibalism, males are passive compare to other species that would try to escape. The idea of "paternal investment" supports the concept of female choice because female spiders will consume males in order to receive an increase in quality offspring. Males may tap into females' sensory bias that may influence females' selection. Male gift-giving spiders are known to provide gifts to females in order to avoid being eaten.

This is a tactic that males may use in order to manipulate females from killing them. Females may have a strong uncontrollable appetite. Males may use this to their advantage by manipulating females through edible gifts.

Counter-adaptation

Traditionally, counter-adaptations are either behavioural or anatomical in order to reduce the traumatic effects implemented by males. Counter-adaptations can be expressed for a variety of species.

Behavioural

Kicking Ability: The Bruchid beetle *Callosobruchus maculatus* is a particular type of species that participate in extra-genitalic traumatic insemination. Male Bruchid beetles have spiky genitals that will puncture the female reproductive tract. Females will perform a "kicking ability" in order to reduce the length of copulation. A study focused on whether inflicting pain on females was beneficial for mating and if females' kicking ability increases their fitness. It was observed that females' who were restricted from kicking actually lowered their fecundity and fitness.

Females that were able to kick had shorter copulation and a reduction of injuries. It was evident that females that were restricted from kicking during the second copulation did not improve the success of sperm completion in the second mate. It was suggested that the spines on the males may act as an anchor to assist them during copulation as a by-product of harming females. Female Bruchid beetle's kicking ability is a counter-adaptation to the spiky genitals of males.

Warding/aggressive Behaviour

Warding and aggressive behaviour can be seen in hermaphrodites as a counter-adaptation. In the context of selecting female roles, it was observed that mates would become aggressive in order to avoid being a female. Species that exhibit this behaviour are marine flatworms and banana slugs.

Marine flat worms *Pseudocerus bifurcus* will perform penis fencing, in which both mates will battle to stab the other one first in order for that individual to take on the female role. Mates will develop an aggressive behaviour when duelling with their mates, but once struck they will display an avoidance behaviour. The aggressive behaviour can be seen as a counter-adaptation because the benefits include a reduction in detrimental effect.

However, the avoidance behaviour displayed by mates who are struck can also be seen as beneficial, because they will be passing on traits that represent a better stabber than themselves (female choice).

Banana slugs *Ariolimax spp.* may display aggressive behaviour in terms of the selection of female roles. Hermaphroditic Banana slugs will either select a male or female role. When banana slugs mate, the female's genital will grip the male resulting in them being locked together. After repeated attempts to break from this lock, the female will bite the male's penis in order to release herself from him (aprophallation). The male will no longer be able to mate as a male. Hermaphrodites are more likely to damage the male's reproductive function compare to the females. This could be a counter-adaptation in order to prevent remating with males without negative consequences on female's fecundity.

Alarm Pheromone

In bed bugs *Cimex lectularius* males will perform traumatic insemination on females. Males may have a counter-adaptation to avoid being pierced by other males. When both sexes are fed, they typically resemble each other, resulting in males accidentally piercing other males. The counter-adaptation is an "alarm pheromone" that acts as an anti-aphrodisiac chemical that is released by males when approached by other males. In addition, males may have specific defence behaviours. Nymphs also release alarm pheromones because they are able to grow to the same size of mature females, resulting in males mistakenly approaching them.

Gift-giving and Faking Death

Sexual cannibalism occurs in gift-giving spiders *Pisaura mirabilis* prior or after copulation. Females are the ones that will kill and consume the male gift-giving spiders. Prior to copulation the male gift-giving spider will come towards the female and provide her with a gift. The female gift-giving spider will acknowledge the gift and begin consuming it. At this point the female is occupied consuming the gift, which allows the male to deposit sperm. During periods of copulation the female will stop eating her gift. As a result the male would fake death (perform thanatosis) to avoid being consumed. Males who perform thanatosis were able to copulate for a longer period. However, a study observed that gift-giving spiders that were able to copulate, usually spent a shorter time because of the fear of being cannibalised.

Males are known to manipulate females through the use of chemical ejaculates that would restrict females from remating. Females may have developed a resistance towards males by consuming their ejaculates. Female resistance could have influenced the purpose of male ejaculates from being used for reproduction to seminal gifts. The shift in roles between males and females could have resulted as females' resistance towards males' aggressive behaviours.

Toxic Spray

The funnel-web spiders *Agelenopsis aperta* exhibits sexual cannibalism. In the event prior or after copulation, female spiders will kill and consume male funnel-web spiders. At the risk of being cannibalised by females, it has been shown that male funnel-web spider has developed a counter-adaptation to prevent being consumed. A study observed that funnel web spiders that were aggressive were swift to assault prey and invaders compared to spiders that were less aggressive. Spiders that are aggressive were more likely to participate in useless non-adaptive killing.

In other words, spiders that are aggressive will kill prey but not consume them. Male funnel web spiders are fearful and cautious when approaching a female in order to engage in copulation. As a result, the male spider has developed a strategy in order to compensate for the female's aggression. The male will move towards the female and spray a toxic substance. The toxic substance will result in the female collapsing and becoming unconscious. The male will reposition the sedative female in order to successfully inseminate her without her waking up and consuming him. Female spiders that are aggressive are more likely to kill prey compared to less aggressive spiders. This reflects how females may have evolved this aggressive behaviour towards males.

Forceful Fighting

Male water striders will perform force copulation by using grasp techniques in order to attach onto the female. This will cause detrimental effects on the females because she is carrying the weight of the male while gliding on the water surface. The female has developed counter-adaptations in order to reduce the traumatic effects of the males.

During force copulation, females may display forceful fighting or intense somersaulting and rolling on the water surface in order to fight off males. The struggle the female endures usually lasts a couple of minutes. The female water strider also has an antigrasping structure known as the connexival spines. A study observed that female water strider *Gerris incognitus* that had longer spines compared to other females were more likely to resist males faster.

Frantic Swimming

Male diving beetles *Dytiscidae* will use grasping devices in order to force females to mate. The female will resist the traumatic effects endured by the male by continuously struggling with him. There has been evidence of counter-adaptation in female diving beetles both behaviourally and anatomically.

Female diving beetles' have what is called an antigrasping structure. Females are seen as polymorphic because this feature is only expressed in some females of the same species. Some females will have morphs with rough dorsal surface while others will have the smooth surface.

The irregular dorsal structures have been shown to decrease male adhesiveness. Although females have a dented dorsal surface, she also has a setose furrows located on her elytra. Darwin viewed that the purpose of female setose furrows was to help males with a better grip during copulation. Even though it appears that females have antigrasping features, it is still insufficient which causes the females to develop the behaviour of "frantic swimming."

When females are at risk of being captured by males, they will swim frantically in order to resist the male. Depending upon the species, some males will have grasping structures such as sucker-shape setae located on their legs in order to attach themselves to the females' structures (pronota or elytra).

Anatomical

There has been evidence of species that developed counter-adaptations towards their mates due to their structure. Some mates will have structures that will reduce traumatic effects in terms of repairing wounds, acting as a disguise or even an evolutionary explanation.

Paragenital Sinuses (Spermalege)

The paragenital sinuses or spermalege can be seen as a "secondary reproductive system." The reason the spermalege has been called a secondary organ, is because it may have evolved as a counter-adaptation for females. The paragenital can consist of layers such as the ectospermalege and the mesospermalege. Some species such as the bed bug and bat bug are composed of a paragenital but the layers can be manifested differently among species.

The ectospermalege is described as an enhancement of the female's abdominal wall. The ectospermalege is usually located where males tend to puncture and may consist of enlarged folds. The mesospermalege is located in the interior surface of the abdominal wall.

The mesospermalege is shaped as a pocket or sac beneath the ectospermalege. The mesospermalege is consumed of haemocytoid cells and collects transmitted ejaculate. The mesospermalege will divert incoming sperm from heading towards the hemocele of the female in addition to being filled with phagocytic hemocytes. The mesospermalege possible rcle could be to destroy sperm. The structure of the paragenital serves as counter-adaptation.

Some benefits of the paragenital sinuses include reducing infliction, wound healing and increase in life span. During traumatic insemination, if a region was wounded, the paragenital would trigger a systemic immune response that would send haemocytes to the wounded area in the paragenital.

The efficiency of the immune response will also respond to pathogens in the same matter. The mesospermalege may allow females to express cryptic choice either directly such as physiology or indirectly by consuming sperm. It was observed that females that experienced puncturing in their spermalege did not affect females compared to females who experienced puncturing outside of the spermalege. As a result those females, who were punctured away from the spermalege, suffered a 50% decrease in egg production. It was compared that puncturing of a female's spermalege was similar to that of non-traumatic copulation because the females were unaffected.

Bed Bug (Cimex Lectularius)

The paragenital may seem beneficial to female bed bugs because they lack aggressive behaviour. In a study, it was observed that during copulation, females would display a passive behaviour rather a forcefully behaviour as seen in the Bruchid beetle. Due to females' lack of resistance, females can actually benefit from having a paragenital that will compensate for their passive behaviour.

In a study artificial insemination was perform in the female's mesospermalege. It was found that females who received inseminations in regions near the mesospermalege faced detrimental effects compare to females who received direct deposit in the mesospermalege. Although this is a well-defined counter-adaptation, it is possible for female bed bugs to develop other methods of resistance.

A study that focused on the traumatic insemination of bed bugs *Cimex lectularius*, observed that when male bed bugs were allowed to mate freely with both fed genders, males were seen to mount both genders. The assumption was made that when both female and males are fed, that their appearance was rather similar and difficult to distinguish. This explains why males were experiencing mounting by other males, suggesting that males have a preference over the appearance of fed females. The idea was that when females were fed, they were much slower and unable to engage in "refusal posture", compare to unfed females who were able to perform this behaviour to mask their ectospermalege. It is possible that females may have other counter-adaptation besides the paragenital which may include a behavioural response and sexual mimicry. Females can manipulate males, by overindulging in food in order to display a male appearance to reduce the risk of traumatic insemination. A behavioural response that females may use to their advantage would be the "refusal posture" which would be a strategy in order to cover her ectospermalege.

African Bat Bug (Afrocimex Constrictus)

The uniqueness of African bat bugs is that both genders are at risk of suffering traumatic insemination by males. Male African bat bug have a paragenital sinuses (morph) that has a "female-like" appearance causing it to be more susceptible to other males. Females express polymorphism in which females may have two types of paragenital sinuses within the same species. For example females who have an open morph will have an appearance that is similar to males, while females who have a closed morph will have a distinct female appearance. Due to this conflict, both genders are equipped with paragenital sinuses. The evolution of paragenital sinuses in both genders can be seen as counter-adaptations.

The paragenital can consist of a mesospermalege or an ectospermalege. In African bat bug, both genders will display the ectospermalege structure, but females in addition to having an ectospermalege, will also have a mesospermalege. The mesospermalege in females is limited to the diffusion of mass cells. Male bat bugs that suffered from traumatic insemination by other males had sperm from other males in their ectospermalege but the sperm was nowhere near the reproductive tract. It can be reasoned that both genders have paragenital in order to reduce the traumatic effects of males. A possible reason for females having both mesospermalege and ectospermalege, compare to males that only have an ectospermalege could be explained by their functions. The mesospermalege will redirect sperm, while ectospermalege is the outer layer that protects piercing. Therefore, females would need to have a mesospermalege in order to properly direct sperm, while males do not because they are not equipped to reproduce. From an evolutionary perspective, females may have cryptic choice due to their mesospermalege.

Sexual Mimicry

Sexual mimicry commonly occurs in various species. Female mimicking males have been a mechanism implemented in order to decrease the likelihood of male harassment. For example, female fiddler crabs performing male-like behaviours. Another example is gravid female lizards that would develop similar male colouration of the body. In particular circumstances, members of both genders would benefit by masking their sexual identity resulting in decreasing the risks of intra and intersexual relations.

Sexual mimicry can provide benefits for either sex, and can exist in a variety of species such as the African bat bug. In African bat bug *Afrocimex constrictus* both genders are at risk of traumatic insemination. Since the male African bat bug displays genitals of a "female-like" appearance, it makes him more susceptible of being punctured by other males by mistaking him as a female.

Females within the same species may display two types of genitals. Out of the two types of genitals, the first genital has similar appearance to males while the other type has more a distinct female appearance. Females who have an open morph or paragenital sinuses most closely resembles that of a male's morph, while females who have a closed morph are distinctly seen as females. It is evident that females who have an open morph of a "male-like" appearance are less likely to suffer traumatic insemination compared to males and closed morph females. It is assumed that open morph females' counter-adaptation would be a sexual mimicry because they are disguising or masking their true identity in order to reduce the detrimental effects implemented by males.

Genital Pads

Female Malabar ricefish are known to experience a form of harassment leading up to forced copulation by males. The male will insert his gonopodium organ near female's genital and allowing the spermatophore to release sperm that will travel into the genital opening of the female. As a result, female Malabar ricefish developed a counter-adaptation.

Female Malabar ricefish may express two forms of counter-adaptation. The first type of counter-adaptation is behavioural in which the female will become very aggressive towards males that are harassing her. The second counter-adaptation is anatomical, in which females have a "genital pad". The purpose of the "genital pad" is to cover the female's genital opening. The characteristics of the "genital pad" are that the skin texture is hard and thick. When males insert their spermatophore into females, injuries were reduced because of the "genital pad". The "genital pad" protect females from injures of the forceful strokes of the gonopodium.

Love Dart

A love dart (also known as a gypsobelum) is a hard, long, sharp, calcareous or chitinous dart which some hermaphroditic land snails and slugs create. Love darts are made in sexually mature animals only, and are used as part of the sequence of events during courtship, before actual mating takes place. Darts are quite large compared to the size of the animal: in the case of the semi-slug genus *Parmarion*, the length of a dart can be up to one fifth that of the semi-slug's foot.

Prior to copulation, each of the two snails (or slugs) attempt to "shoot" one (or more) darts into the other snail (or slug). There is no organ to receive the dart; this action is more analogous to a stabbing, or to being shot with an arrow. The dart does not fly through the air to reach its target however; instead it is fired as a contact shot.

The love dart is emphatically not a penial stylet (in other words this is *not* an accessory organ for sperm transfer). The exchange of sperm between both of the two land snails is a completely separate part of the mating progression. Nevertheless, recent research shows that use of the dart can strongly favour the reproductive outcome for the snail that is able to lodge a dart first in its partner. This is because mucus on the dart introduces a hormone-like substance that allows far more of its sperm to survive.

Love darts, also known as shooting darts, or just as darts, are shaped in many distinctive ways which vary considerably between species. What all the shapes of love darts have in common is their harpoon-like or needle-like ability to pierce.

The Mating Dance

Figure 4: Courtship in the edible snail, Helix pomatia

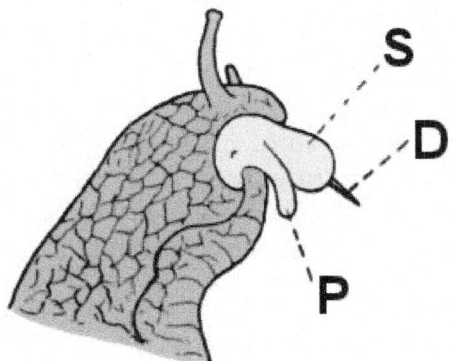

Figure 5: Drawing of the head of a *Helix pomatia* prior to mating, showing the everted penis, and the dart sac in the process of shooting a love dart.

S - dart sac (*bursa telae*)

D - love dart

P - penis

Mating begins with a courting ritual. For example, in land snails of the genus *Helix*, including the escargot *Helix pomatia*, and the common garden snail *Helix aspersa* (also known as *Cornu aspersa* and *Cantareus aspersus*), copulation is preceded by an elaborate tactile courtship.

The two snails circle around each other for up to six hours, touching with their tentacles, and biting lips and the area of the genital pore, which shows some preliminary signs of the eversion of the penis. As the snails approach mating, hydraulic pressure builds up in the blood sinus surrounding the organ housing the dart. Each snail manoeuvres to get its genital pore in the best position, close to the other snail's body. Then, when the body of one snail touches the other snail's genital pore, it triggers the firing of the dart.

The darting can sometimes be so forceful that the dart ends up buried in the internal organs. It can also happen that a dart will pierce the body or head entirely, and protrude on the other side.

After both snails have fired their darts, the snails copulate and exchange sperm.

A snail does not have a dart to fire the very first time it mates, because the first mating is necessary to trigger the process of dart formation. Once a snail has mated, it fires a dart before some, but not all, subsequent matings. A snail often mates without having a dart to use, because it takes time to create a replacement dart. In the case of the garden snail *Helix aspersa*, it takes a week for a new dart to form.

The dart is shot with some variation in force, and with considerable inaccuracy, such that one-third of the darts that are fired in *Helix aspersa* either fail to penetrate the skin, or miss the target altogether. Snails have only very simple visual systems and cannot see well enough to use vision to help aim the darts.

Function

Although the existence and use of love darts in snails has been known for at least several centuries, until recently the actual function of love darts was not properly understood.

It was long assumed that the darts had some sort of "stimulating" function, and served to make copulation more likely. It was also suggested that darts might be a "gift" of calcium. These theories have proved not to be accurate; recent research has led to a better understanding of the strange phenomenon of love darts.

The two species that have been studied the most are *Helix aspersa*, the garden snail, and *Helix pomatia*, the edible escargot. In *Helix aspersa* (aka *Cantareus aspersus*), the dart is coated with a special mucus which contains a hormone-like substance. This substance contracts one part of the female half of the reproductive system of the snail that is struck with the dart, and this allows many more sperm to survive, significantly increasing the likelihood of a successful fertilization.

Morphology of Darts

The love dart, also known as a "gypsobelum", is often made of calcium carbonate which is secreted by a specialised organ within the reproductive system of several families of air-breathing snails and slugs, mainly in terrestrial pulmonate gastropod mollusks within the clade Stylommatophora.

Darts can range in size from about 30 mm long in the larger snail species, down to about 1 mm in the smallest snails that have darts. Typically most darts are less than 5 mm long, but still they are substantial compared with the size of the animal.

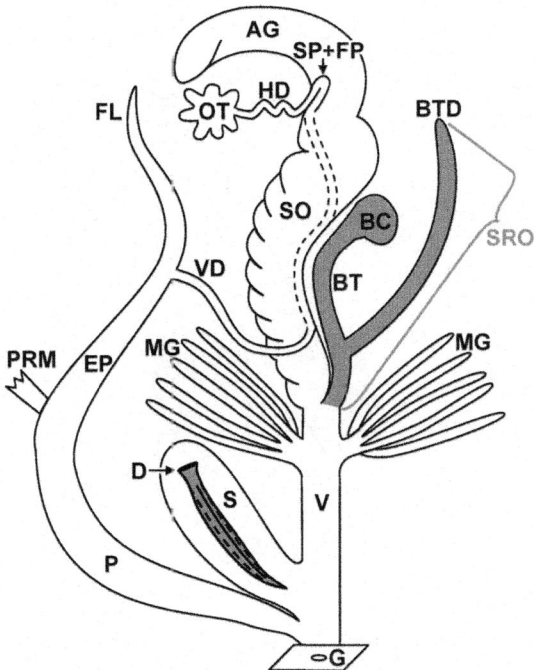

Figure 6: Simplified diagram of the reproductive system of a pulmonate land snail

D = love dart; S = stylophore or dart sac (*bursa telae*); P = penis; V = vagina; G = genital pore; MG = mucus glands

There is considerable variety in both the overall shape and the cross section of the love dart. The morphology (shape and form) of the dart is species-specific. For example, individual snails of the two rather similar helicid species *Cepaea hortensis* and *Cepaea nemoralis* can sometimes only be distinguished by examining the shape of the love dart and the vaginal mucus glands, which in the anatomical diagram are marked "MG" and are positioned off of the structure marked "V".

Anatomical Context

There is a complex hermaphroditic reproductive system in pulmonate snails (those snails that have a lung rather than a gill or gills.) Their reproductive system is completely internal, except for the active protrusion (eversion) cf the penis for copulation. The outer opening of the reproductive system is called the "genital pore"; it is positioned on the right hand side, very close to the head of the animal. This opening is virtually invisible however, unless it is actively in use.

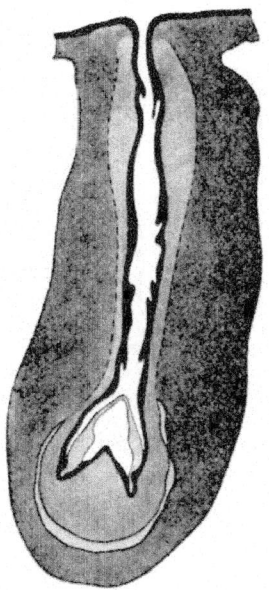

Figure 7: Drawing of a transverse section of the dart sac (also known as the *bursa telae*) of *Helix pomatia* during the process of creating a new love dart.

The love dart is created and stored before use in a highly muscular internal anatomical structure known as the *stylophore* or dart sac (also known as the *bursa telae*). The exact positioning of the stylophore varies, but it is in the vicinity of the eversible penis and the vagina, where these two structures open into the "atrium", a common area right inside the genital pore.

The opening of the stylophore leads directly into the atrium in certain species in the families Vitrinidae, Parmacellidae, Helminthoglyptidae, Bradybaenidae, Urocyclidae, Ariophantidae, and Dyakiidae.

The opening of the stylophore can instead lead to the penis, as is the case in some species of Aneitinae (a subfamily of Athoracophoridae), Sagdidae, Euconulidae, Gastrodontidae and Onchidiidae. Alternatively, it can lead to the vagina, as in the case in some species of Ariopeltinae (a subfamily of Oopeltidae), Ariolimacinae (a subfamily of Ariolimacidae), Philomycidae, other species within the Bradybaenidae, and also in the Hygromiidae, Helicidae and Dyakiidae.

Only two families have darts present in every species: the Bradybaenidae and in the Dyakiidae. In all the other families there is reduction or loss of dart-making ability in some of the species (cf.).

Many species have only one dart sac, however other species have several. Snails in the family Bradybaenidae have more than one dart sac, and some species of Hygromiidae and Helmintoglyptidae have four dart sacs. Some Urocyclidae have up to 70 darts.

Occurrence Within the Pulmonate Snails and Slugs

All pulmonate land snails are hermaphrodites, and have a complete and rather elaborate set of both male and female reproductive organs, but the majority of pulmonate land snails have no love darts and no dart sac.

Calcareous Darts

Calcareous (composed of calcium carbonate) darts are found in a limited number of pulmonate families within the Stylommatophora.

Most of these families are within the land snail superfamily Helicoidea: Helicidae, Bradybaenidae, Helminthoglyptidae, Hygromiidae, Humboldtianidae (previously considered to be a part of the Hygromiidae).

Calcium carbonate darts are also found in the family Zonitidae within the superfamily Zonitoidea, and in one family of slugs, the Philomycidae, which are within the superfamly Arionoidea.

Lightly calcified darts occur in the snail and semi-slug family Urocyclidae, within the superfamily Helicarionoidea.

Chitinous Darts

Chitinous (composed of chitin) love darts occur in the pulmonate land snail families Ariophantidae (superfamily Helicarionoidea), in the family Helicarionidae (superfamily Helicarionoidea), in the Vitrinidae (superfamily Limacoidea), and in the slug family Parmacellidae (superfamily Parmacelloidea).

Within the more ancient clade Systellommatophora, chitin darts are found in the pulmonate sea slugs of the family Onchidiidae, in the superfamily Onchidioidea.

Cartilaginous Darts

Love darts made of cartilage occur in the family Gastrodontidae.

Evolution of Love Darts

Because of the presence of darts in many superfamilies of the Stylommatophora, it seems likely that love darts appeared during the early evolution of the Pulmonata, and that the ancestors of the Stylommatophora possessed darts already.

During evolution, darts appear to have been lost secondarily, i.e., after they had evolved and been functional. Vestigial darts (ones that exist only in a rudimentary condition) occur in the family Sagdidae., and in many Helicoidea, the surrounding organs have also degenerated (become non-functional). The *sarcobelum* is a fleshy or cuticle-coated papilla which is considered to be a degenerated, previously dart-bearing, organ.

Species Variability

Love darts are shaped in many distinctive ways, and vary considerably between species. The morphology of the dart is almost always species-specific.

Some darts have a round cross section, others are bladed or vaned. In some cases the blades on the sides of the dart are bifurcated or divided into two parts. Some darts are shaped like a needle or a thorn, others have a tip like an arrowhead, or look like a dagger. What all the shapes have in common is their ability to pierce.

Notes on Specific Species

Bonobos: The Bonobo, which has a matriarchal society, is a fully bisexual species — both males and females engage in sexual behaviour with the same and the opposite sex, with females being particularly noted for engaging in sexual behaviour with each other and at up to

75% of sexual activity being nonreproductive. Primatologist Frans de Waal believes that Bonobos use sexual activity to resolve conflict between individuals. Sexual activity occurs between almost all ages and sexes of Bonobo societies.

Birds

Some black swans of Australia form sexually active male-male mated pairs and steal nests, or form temporary threesomes with females to obtain eggs, driving away the female after she lays the eggs. More of their cygnets survive to adulthood than those of different-sex pairs possibly due to their superior ability to defend large portions of land.

In early February 2004 the *New York Times* reported that a male pair of chinstrap penguins named Roy and Silo in the Central Park Zoo in New York City were partnered and had successfully hatched a female chick from an egg. Other penguins in New York have also been reported to be forming same-sex pairs.

Zoos in Japan and Germany have also documented male penguin couples. The couples have been shown to build nests together and use a stone to replace an egg in the nest. Researchers at Rikkyo University in Tokyo, found twenty such pairs at sixteen major aquariums and zoos in Japan. Bremerhaven Zoo in Germany attempted to break up the male couples by importing female penguins from Sweden and separating the male couples; they were unsuccessful. The zoo director stated the relationships were too strong between the couples.

Recently, a mated pair of swans in Boston were found to both be female. They too had attempted to raise eggs together.

Studies have shown that ten to fifteen percent of female western gulls in some populations in the wild prefer other females.

As many as 19% of Mallard pairs in a given population have been observed to consist of male-male homosexuals.

Lizards

Whip-tailed lizard females have the ability to reproduce through parthenogenesis and as such males are rare and sexual breeding non-standard. Females engage in sexual behaviour to stimulate ovulation, with their behaviour following their hormonal cycles; during low levels of estrogen, these (female) lizards engage in "masculine" sexual roles. Those animals with currently high estrogen levels assume "feminine" sexual roles.

Lizards that perform the courtship ritual have greater fecundity than those kept in isolation due to an increase in hormones triggered by the sexual behaviours. So, even though asexual whiptail lizards populations lack males, sexual stimuli still increase reproductive success.

From an evolutionary standpoint these females are passing their full genetic code to all of their offspring rather than the 50% of genes that would be passed in sexual reproduction. Certain species of gecko also reproduce by parthenogenesis.

Flatworm

Penis fencing is a mating behaviour engaged in by certain species of flatworm, such as *Pseudobiceros bedfordi*. Species which engage in the practice are hermaphroditic, possessing both eggs and sperm-producing testes.

The species "fence" using two-headed dagger-like penises which are pointed, and white in colour. One organism inseminates the other. The sperm is absorbed through pores in the skin, causing fertilization.

Sheep

An October 2003, study by Dr. Charles E. Roselli et al. (Oregon Health & Science University) states that homosexuality in male sheep (found in eight percent of rams) is associated with a region in the rams' brains which the authors call the "ovine Sexually Dimorphic Nucleus" (oSDN) which is half the size of the corresponding region in other male sheep.

However, some view this study to be flawed in that the determination of homosexuality within the sheep, (sample population of twenty-seven for the study), was to have animals who were unable to mount female ewes placed in a cage with two stanchioned males and two unstanchioned females (that is, the males could not move or struggle while the females could). Given the aggressive nature of the sheep copulation, the uneven treatment of males and females, many see this as simply evidence that the sheep in question were unable to be aggressive enough to mount females. Some say that the results were situational sexuality, unlike the bonds seen in human homosexuality.

The scientists found that, "The oSDN in rams that preferred females was significantly larger and contained more neurons than in male-oriented rams and ewes. In addition, the oSDN of the female-oriented rams expressed higher levels of aromatase, a substance that converts testosterone to estradiol, an estrogen hormone believed to facilitate typical male sexual behaviours. Aromatase expression was no different between male-oriented rams and ewes."

"The dense cluster of neurons that comprise the oSDN express cytochrome P450 aromatase. Aromatase mRNA levels in the oSDN were significantly greater in female-oriented rams than in ewes, whereas male-oriented rams exhibited intermediate levels of expression." These results suggest that "...naturally occurring variations in sexual partner preferences may be related to differences in brain anatomy and its capacity for estrogen synthesis." As noted previously, given the potential unagressiveness of the male population in question, the differing aromatase levels may also have been evidence of aggression levels, not sexuality. The results of this study have not been confirmed by others.

Spotted Hyena

The female Spotted Hyena has a unique urinary-genital system, closely resembling the penis of the male, called a pseudo-penis. The family structure is matriarchal and dominance relationships with strong sexual elements are routinely observed between related females.

They are notable for using visible sexual arousal as a sign of *submission* and not *dominance*, in males as well as females (females have a sizable erectile clitoris), to the extent that biologist Robert Sapolsky speculates that in order to facilitate this, their sympathetic and parasympathetic nervous systems may be partially reversed in respect to their reproductive organs.

Bottlenose Dolphins

Bottlenose Dolphin males have been observed working in pairs to follow and/or restrict the movement of a female for weeks at a time, waiting for her to become sexually receptive. The same pairs have also been observed engaging in intense sexual play with each other.

Janet Mann, a professor of biology and psychology at Georgetown University, argues that the common same-sex behaviour among male dolphin calves is about bond formation and benefits the species evolutionarily. They cite studies that have shown the dolphins later in life are bisexual and the male bonds forged from homosexuality work for protection as well as locating females with which to reproduce.

In 1991 an English man was prosecuted for allegedly having sexual contact with a dolphin. The man was found not guilty after it was revealed at trial that the dolphin was known to tow bathers through the water by hooking its large penis around them.

Seahorses

Seahorses, long upheld as monogamous and mating for life, are identified as "promiscuous, flighty, and more than a little bit gay" according to research published in 2007.

Scientists at 15 aquariums studied 90 seahorses of 3 species. Of 3168 sexual encounters, 37% were same sex acts. Flirting was common (up to 25 potential partners a day of both genders); only one species (the British Spiny Seahorse) included faithful representatives, and for these 5 of 17 were faithful, 12 were not. Bisexuality was widespread and considered "both a great surprise and a shock", with big-bellied seahorses of both genders not showing partner preference. 1986 contacts were male-female, 836 were female-female and 346 were male-male.

Lions

Male lions often lead their social groups jointly with one or more of their brothers. To ensure loyalty, the male co-leaders will "strengthen the bonds" by often having sex with each other.

Horses

Anecdotal evidence suggest that some horses have environment or appearance preferences when selecting mates. There is also anecdotal evidence of limited bisexual behaviour in some stallions, although there is (as of 2008) no conclusive scientific confirmation. The anecdotal evidence claims this is most likely to occur in a single isolated group, with no access to mares.

Humboldt Penguins

In 2009 at a zoo in Bremerhaven, Germany, two male adult humboldt penguins adopted an egg that had been abandoned by its biological parents. After the egg hatched, the two penguins raised, protected, cared for, and fed the chick in the same manner that heterosexual penguins raise their own biological offspring.

Other Evidence of Interspecies Sexual Activity

Looking back in history, current research into human evolution tends to confirm that in some cases, interspecies sexual activity may have been responsible for the evolution of entire new species. Analysis of human and animal genes in 2006 provides strong evidence that after humans had diverged from other apes, interspecies mating nonetheless occurred regularly enough to change certain genes in the new gene pool:

A new comparison of the human and chimp genomes suggests that after the two lineages separated, they may have begun interbreeding. A principal finding is that the X chromosomes of humans and chimps appear to have diverged about 1.2 million years more recently than the other chromosomes.

The research suggests that:

> *There were in fact two splits between the human and chimp lineages, with the first being followed by interbreeding between the two populations and then a second split. The suggestion of a hybridisation has startled paleoanthropologists, who nonetheless are "treating the new genetic data seriously."*

> *The* Washington Post *comments, "If this theory proves correct, it will mean modern people are descended from something akin to chimp-human hybrids."*

Role in Discussion of Human Eexuality

Information about animal sexuality frequently arises as a persuasive device in arguments regarding human sexuality. Originally, the lack of documented animal sexual behaviour deviant from heterosexual sexual monogamy was used to argue that the dominant heterosexual monogamy of most modern human societies is more natural and acceptable. Likewise, the lack of documented sex between animals for the purpose of pleasure was used to promote the moral standard of reserving sex primarily for procreation. Proponents of alternate sexuality attribute this early lack of documented evidence to an observer bias in researchers, who, they argue, tended to interpret sexual behaviour inconsistent with their values as other behaviour.

With increasing published evidence of different types of sexual behaviour between animals, arguments for heterosexual monogamy in human society have moved towards characterising these behaviours as resulting from differences between humans and animals, and in particular on ambiguity in motivation and subjective experience in animals, which is difficult to study. Arguments identifying human and animal behaviour are characterised as anthropomorphism, and in some cases an opposite observer bias is attributed to researchers. Supporters of alternate sexuality embrace the new research as confirmation of the naturalness of alternate sexual behaviour and evidence of its long-term feasibility and utility. In both cases, any argument that claims that something is good or right because it is natural, or that something is bad or wrong because it is unnatural or artificial is known as the appeal to nature fallacy.

Freemartin

A freemartin or free-martin (sometimes martin heifer) is an infertile female mammal which has masculinised behaviour and non-functioning ovaries. Genetically and externally the animal is female, but it is sterilised in the womb by hormones from a male twin, becoming an infertile partial intersex. Freemartinism is the normal outcome of mixed-sex twins in all cattle species that have been studied, and it also occurs occasionally in other mammals including sheep, goats and pigs.

The 18th-century physician John Hunter discovered that a freemartin always has a male twin. It was hypothesized early in the 20th century that masculinising factors travel from the male twin to the female twin through the vascular connections of the placenta because of the vascular fusion and affect the internal anatomy of the female. Several researchers made the discovery that a freemartin results when a female fetus has its chorion fuse in the uterus with that of a male twin. The result was published in 1916 by Tandler and Keller. The discovery was made independently by American biologist Frank R. Lillie, who published it in Science in 1916. Both teams are now credited with the discovery. In rural areas folklore often claimed this condition was not just peculiar to cattle, but extended also to human twins; this belief perpetuated for generations, as was mentioned in the writings of Bede.

Mechanism

In most cattle twins, the blood vessels in the chorions become interconnected, allowing blood from each twin to flow around the other. If both fetuses are the same sex this is of no significance, but if they are different, male hormones pass from the male twin to the female twin. The male hormones then masculinize the female twin, and the result is a freemartin. The degree of masculinization is greater if the fusion occurs earlier in the pregnancy – in about ten percent of cases no fusion takes place and the female remains fertile. The male twin is largely unaffected by the fusion, although the size of the testicles may be slightly reduced. Testicle size is associated with fertility, so there may be some reduction in bull fertility.

Freemartins behave and grow in a similar way to castrated male cattle (steers).

Diagnosis

If suspected, a test can be done to detect the presence of the male Y-chromosomes in some circulating white blood cells of the subject. Genetic testing for the Y-chromosome can be performed within days of birth and can aid in the early identification of a sterile female bovine. Physical examination of the calf may also reveal differences: many (but not all) freemartins have a short vagina compared with that of a fertile heifer.

Other Animals

A freemartin is the normal outcome of mixed twins in all cattle species which have been studied. It does not normally occur in most other mammals, though it has been recorded in sheep, goats, and pigs.

Uses

Freemartins are occasionally used in stem cell and immunology research. During fetal development cells are exchanged between the fused circulations of the bovine twins. Up to 95% of the freemartin's blood cells can be derived from those of its twin brother. Male-derived cells and their progeny can be easily visualised in the freemartin tissues, as only they contain the male Y chromosome. Thus, by analysing these tissues, one is able to investigate the capacity of hematopoietic stem cells or other circulating cells to produce other tissues in addition to blood. The freemartin model allows one to analyse perfectly healthy and unmanipulated animals, without resorting to transplantation often used in stem cell research.

Pair Bond

In biology, a pair bond is the strong affinity that develops in some species between the males and females in a pair, potentially leading to breeding. Pair-bonding is a term coined in the 1940s that is frequently used in sociobiology and evolutionary psychology circles. The term often implies either a lifelong socially monogamous relationship or a stage of mating interaction in socially monogamous species. It is sometimes used in reference to human relationships.

Monogamous voles, such as prairie voles, have significant differences in the density and distribution of vasopressin receptors in their brain when compared to polygamous voles. These differences are located in the ventral forebrain and the dopamine-mediated reward pathway. Both vasopressin and dopamine act in this region to coordinate rewarding activities such as mating, and regulate selective affiliation. These species specific differences have shown to correlate with social analyss, and in monogamous prairie voles are important for facilitation of pair bonding.

Varieties

According to evolutionary psychologists David Barash and Judith Lipton, from their 2001 book *The Myth of Monogamy*, there are several varieties of pair bonds:

- *Short-term pair-bond:* a transient mating or associations
- *Long-term pair-bond:* bonded for a significant portion of the life cycle of that pair
- *Lifelong pair-bond:* mated for life
- *Social pair-bond:* attachments for territorial or social reasons, as in cuckold situations
- *Clandestine pair-bond:* quick extra-pair copulations
- *Dynamic pair-bond:* e.g. gibbon mating systems being analogous to "swingers".

Examples

When discussing the social life of the bank swallow, Lipton and Barash state:

> *For about four days immediately prior to egg-laying, when copulations lead to fertilizations, the male bank swallow is very busy, attentively guarding his female. Before this time, as well as after—that is, when her eggs are not ripe, and again after his genes are safely tucked away inside the shells—he goes seeking extra-pair copulations with the mates of other males...who, of course, are busy with defensive mate-guarding of their own.*

In various species, males provide parental care and females mate with multiple males. For example, recent studies show that extra-pair copulation frequently occurs in monogamous birds in which a "social" father provides intensive care for its "social" offspring.

A University of Florida scientist reports that male sand gobies work harder at building nests and taking care of eggs when females are present – the first time such "courtship parental care" has been documented in any species.

Chapter 5

Model Organism

A model organism is a non-human species that is extensively studied to understand particular biological phenomena, with the expectation that discoveries made in the organism model will provide insight into the workings of other organisms. Model organisms are widely used to explore potential causes and treatments for human disease when human experimentation would be unfeasible or considered less ethical. This strategy is made possible by the common descent of all living organisms, and the conservation of metabolic and developmental pathways and genetic material over the course of evolution. Studying model organisms can be informative, but care must be taken when generalising from one organism to another.

Selecting a Model Organism

Models are those organisms with a wealth of biological data that make them attractive to study as examples for other species and/or natural phenomena that are more difficult to study directly. Continual research on these organisms focus on a wide variety of experimental techniques and goals from many different levels of biology—from ecology, behaviour, and biomechanics, down to the tiny functional scale of individual tissues, organelles, and proteins. Inquiries about the DNA of organisms are classed as genetic models (with short generation times, such as the fruitfly and nematode worm), experimental models, and genomic parsimony models, investigating pivotal position in the evolutionary tree. Historically, model organisms include a handful of species with extensive genomic research data, such as the NIH model organisms.

Often, model organisms are chosen on the basis that they are amenable to experimental manipulation. This usually will include characteristics such as short life-cycle, techniques for genetic manipulation (inbred strains, stem cell lines, and methods of transformation) and non-specialist living requirements. Sometimes, the genome arrangement facilitates the sequencing of the model organism's genome, for example, by being very compact or having a low proportion of junk DNA (e.g. yeast, *Arabidopsis*, or pufferfish).

When researchers look for an organism to use in their studies, they look for several traits. Among these are size, generation time, accessibility, manipulation, genetics, conservation of mechanisms, and potential economic benefit. As comparative molecular biology has become more common, some researchers have sought model organisms from a wider assortment of lineages on the tree of life.

Use of Model Organisms

There are many model organisms. One of the first model systems for molecular biology was the bacterium *Escherichia coli*, a common constituent of the human digestive system. Several of the bacterial viruses (bacteriophage) that infect *E. coli* also have been very useful for the study of gene structure and gene regulation (e.g. phages Lambda and T4). However, bacteriophages are not organisms because they lack metabolism and depend on functions of the host cells for propagation.

In eukaryotes, several yeasts, particularly *Saccharomyces cerevisiae* ("baker's" or "budding" yeast), have been widely used in genetics and cell biology, largely because they are quick and easy to grow. The cell cycle in a simple yeast is very similar to the cell cycle in humans and is regulated by homologous proteins.

The fruit fly *Drosophila melanogaster* is studied, again, because it is easy to grow for an animal, has various visible congenital traits and has a polytene (giant) chromosome in its salivary glands that can be examined under a light microscope. The roundworm *Caenorhabditis elegans* is studied because it has very defined development patterns involving fixed numbers of cells, and it can be rapidly assayed for abnormalities.

Important Model Organisms

Viruses include:

* Phage Lambda
* Phi X 174 - its genome was the first ever to be sequenced. The genome is a circle of 11 genes, 5386 base pairs in length.
* Tobacco mosaic virus.

Prokaryotes

Prokaryotes include:

- *Escherichia coli* (*E. coli*) - This common, Gram-negative gut bacterium is the most widely-used organism in molecular genetics.

- *Bacillus subtilis* - an endospore forming Gram-positive bacterium

- *Caulobacter crescentus* - a bacterium that divides into two distinct cells used to study cellular differentiation.

- *Mycoplasma genitalium* - a minimal organism

- *Vibrio fischeri* - quorum sensing, bioluminescence and animal-bacterial symbiosis with Hawaiian Bobtail Squid

- *Synechocystis*, a photosynthetic cyanobacterium widely used in photosynthesis research.

- *Pseudomonas fluorescens*, a soil bacterium that readily diversifies into different strains in the lab.

Eukaryotes

Eukaryotes include:

- Protists:

- *Chlamydomonas reinhardtii* - a unicellular green alga used to study photosynthesis, flagella and motility, regulation of metabolism, cell-cell recognition and adhesion, response to nutrient deprivation and many other topics. *Chlamydomonas reinhardtii* has a well-studied genetics, with many known and mapped mutants and expressed sequence tags, and there are advanced methods for genetic transformation and selection of genes. Sequencing of the *Chlamydomonas reinhardtii* genome was reported in October 2007. A *Chlamydomonas* genetic stock centre exists at Duke University, and an international *Chlamydomonas* research interest group meets on a regular basis to discuss research results. *Chlamydomonas* is easy to grow on an inexpensive defined medium.

- *Dictyostelium discoideum* is used in molecular biology and genetics (its genome has been sequenced), and is studied as an example of cell communication, differentiation, and programmed cell death.

- *Tetrahymena thermophila* - a free living freshwater ciliate protozoan.
- *Emiliania huxleyi* - a unicellular marine coccolithophore alga, extensively studied as a model phytoplankton species.
- *Thalassiosira pseudonana* - a unicellular marine diatom alga, extensively studied as a model marine diatom since its genome was published in 2004.

Fungi

- *Ashbya gossypii*, cotton pathogen, subject of genetics studies (polarity, cell cycle)
- *Aspergillus nidulans*, mold subject of genetics studies
- *Coprinus cinereus*, mushroom (genetic studies of mushroom development, genetic studies of meiosis)
- *Neurospora crassa* - orange bread mold (genetic studies of meiosis, metabolic regulation, and *circadian rhythm*)
- *Saccharomyces cerevisiae*, baker's yeast or budding yeast (used in brewing and baking)
- *Schizophyllum commune* - model for mushroom formation.
- *Schizosaccharomyces pombe*, fission yeast, (cell cycle, cell polarity, RNAi, centromere structure and function, transcription)
- *Ustilago maydis*, dimorphic yeast and plant pathogen of maize (dimorphism, plant pathogen, transcription).

Plants

- *Arabidopsis thaliana*, currently the most popular model plant. This herbaceous dicot belonging to Brassicaceae family is a plant closely related to the mustard plant. Its small stature and short generation time facilitates rapid genetic studies, and many phenotypic and biochemical mutants have been mapped. Arabidopsis was the first plant to have its genome sequenced. Its genome sequence, along with a wide range of information concerning *Arabidopsis*, is maintained by the TAIR database.

  (Plant physiology, Developmental biology, Molecular genetics, Population genetics, Cytology, Molecular biology)
- *Selaginella moellendorffii* is a remnant of an ancient lineage of vascular plants and key to understanding the evolution of

land plants. It has a small genome size (~110Mb) and its sequence was released by the Joint Genome Institute in early 2008. *(Evolutionary biology, Molecular biology)*

*   *Brachypodium distachyon* is an emerging experimental model grass that has many attributes that make it an excellent model for temperate cereals. *(Agronomy, Molecular biology, Genetics)*
*   *Lotus japonicus* a model legume used to study the symbiosis responsible for nitrogen fixation. *(Agronomy, Molecular biology)*.

Lemna Gibba

*   *Lemna gibba* is a rapidly-growing aquatic monocot, one of the smallest flowering plants. Lemna growth assays are used to evaluate the toxicity of chemicals to plants in ecotoxicology. Because it can be grown in pure culture, microbial action can be excluded. Lemna is being used as a recombinant expression system for economical production of complex biopharmaceuticals. It is also used in education to demonstrate population growth curves.

Zea Mays

*   Maize (*Zea mays* L.) is a cereal grain. It is a diploid monocot with 10 large chromosome pairs, easily studied with the microscope. Its genetic features, including many known and mapped phenotypic mutants and a large number of progeny per cross (typically 100-200) facilitated the discovery of transposons ("jumping genes"). Many DNA markers have been mapped and the genome has been sequenced. *(Genetics, Molecular biology, Agronomy)*
*   *Medicago truncatula* is a model legume, closely related to the common alfalfa. Its rather small genome is currently being sequenced. It is used to study the symbiosis responsible for nitrogen fixation. *(Agronomy, Molecular biology)*
*   *Mimulus* is a model organism used in evolutionary and functional genomes studies. This specie pertain to Phrymaceae family, with ca. 120 species. Several genetic resources has been designed for the study of this genera, some are free access.
*   *Tobacco BY-2 cells* is suspension cell line from tobacco *(Nicotiana tabaccum)*. Useful for general plant physiology

studies on cell level. Genome of this particular cultivar will be not sequenced (at least in near future), but sequencing of its wild species *Nicotiana tabaccum* is presently in progress. *(Cytology, Plant physiology, Biotechnology)*

- Rice *(Oryza sativa)* is used as a model for cereal biology. It has one of the smallest genomes of any cereal species, and sequencing of its genome is finished. *(Agronomy, Molecular biology)*

Physcomitrella Patens

- *Physcomitrella patens* is a moss increasingly used for studies on development and molecular evolution of plants. It is so far the only non-vascular plant(and so the only "primitive" plant) with its genome completely sequenced. Moreover, it is currently the only land plant with efficient gene targeting that enables gene knockout. The resulting knockout mosses are stored and distributed by the International Moss Stock Centre. *(Plant physiology, Evolutionary biology, Molecular genetics, Molecular biology)*

- *Populus* is a genus used as a model in forest genetics and woody plant studies. It has a small genome size, grows very rapidly, and is easily transformed. The genome sequence of Poplar (Populus trichocarpa) sequence is publicly available.

Animals

Invertebrates:

- *Amphimedon queenslandica*, a demosponge from the phylum Porifera used as a model for evolutionary developmental biology and comparative genomics

- *Arbacia punctulata*, the purple-spined sea urchin, classical subject of embryological studies

- *Aplysia*, a sea slug, whose ink release response serves as a model in neurobiology and whose growth cones serve as a model of cytoskeletal rearrangements

- *Branchiostoma floridae*, a species commonly known as amphioxus or lancelet from the subphylum Cephalochordata of the phylum Chordata used as a model for understanding the evolution of nonchordate deuterostomes, invertebrate chordates, and vertebrates

- *Caenorhabditis elegans*, a nematode, usually called *C. elegans* - an excellent model for understanding the genetic control of development and physiology. *C. elegans* was the first multicellular organism whose genome was completely sequenced

- *Ciona intestinalis*, a sea squirt

- *Drosophila*, usually the species *Drosophila melanogaster* - a kind of fruit fly, famous as the subject of genetics experiments by Thomas Hunt Morgan and others. Easily raised in lab, rapid generations, mutations easily induced, many observable mutations. Recently, Drosophila has been used for neuropharmacological research. *(Molecular genetics, Population genetics, Developmental biology)*.

- *Euprymna scolopes,* the Hawaiian bobtail squid, model for animal-bacterial symbiosis, bioluminescent vibrios

- *Hydra (genus)*, a Cnidarian, is the model organism to understand the processes of regeneration and morphogenesis, as well as the evolution of bilaterian body plans

- *Loligo pealei*, a squid, subject of studies of nerve function because of its giant axon (nearly 1 mm diametre, roughly a thousand times larger than typical mammalian axons)

- *Macrostomum lignano*, a free-living, marine flatworm, a model organism for the study of stem cells, regeneration, ageing, gene function, and the evolution of sex. Easily raised in the lab, short generation time, in determined growth, complex behaviour

- *Mnemiopsis leidyi*, from the phylum Ctenophora (comb jelly) used as a model for evolutionary developmental biology and comparative genomics

- *Nematostella vectensis*, a sea anemone from the phylum Cnidaria used as a model for evolutionary developmental biology and comparative genomics

- *Oikopleura dioica*, an appendicularia, a free-swimming tunicate (or urochordate))

- *Oscarella carmela* a homoscleromorph sponge (phylum Porifera) used as a model in evolutionary developmental biology

- *Parhyale hawaiensis* an amphipod crustacean, used in evolutionary developmental (evo-devo) studies, with an extensive toolbox for genetic manipulation.

- *Platynereis dumerilii* a marine polychaetous annelid, which evolved very slowly and therefore retained many ancestral features.
- *Pristionchus pacificus*, a roundworm used in evolutionary developmental biology in comparative analyses with *C. elegans*
- *Schmidtea mediterranea* a freshwater planarian; a model for regeneration and development of tissues such as the brain and germline
- Stomatogastric ganglion of various arthropod species; a model for motor pattern generation seen in all repetitive motions
- *Strongylocentrotus purpuratus*, the purple sea urchin, widely used in developmental biology
- *Symsagittifera roscoffensis*, a flatworm, subject of studies of bilaterian body plan development
- *Tribolium castaneum*, the flour beetle - a small, easily kept darkling beetle used especially in behavioural ecology experiments
- *Trichoplax adhaerens*, a very simple free-living animal from the phylum Placozoa used as a model in evolutionary developmental biology and comparative genomics.

Vertebrates

- Guinea pig (*Cavia porcellus*) - used by Robert Koch and other early bacteriologists as a host for bacterial infections, hence a byword for "laboratory animal" even though less commonly used today
- Chicken (*Gallus gallus domesticus*) - used for developmental studies, as it is an amniote and excellent for micromanipulation (e.g. tissue grafting) and over-expression of gene products
- Cat (*Felis sylvestris catus*) - used in neurophysiological research
- Dog (*Canis lupus familiaris*) - an important respiratory and cardiovascular model, also contributed to the discovery of classical conditioning.
- Hamster - first used to study kala-azar (leishmaniasis)
- Mouse (*Mus musculus*) - the classic model vertebrate. Many inbred strains exist, as well as lines selected for particular traits, often of medical interest, e.g. body size, obesity, muscularity. *(Quantitative genetics, Molecular evolution, Genomics)*

- Lamprey - spinal cord research
- Medaka (*Oryzias latipes*, the Japanese ricefish) - an important model in developmental biology, and has the advantage of being much sturdier than the traditional Zebrafish
- Rat (*Rattus norvegicus*) - particularly useful as a toxicology model; also particularly useful as a neurological model and source of primary cell cultures, owing to the larger size of organs and suborganellar structures relative to the mouse. *(Molecular evolution, Genomics)*
- Rhesus macaque (*Macaca mulatta*) - used for studies on infectious disease and cognition
- Cotton rat (*Sigmodon hispidus*) - formerly used in polio research
- Zebra finch (*Taeniopygia guttata*) - used in the study of the song system of songbirds and the study of non-mammalian auditory systems
- Takifugu (*Takifugu rubripes*, a pufferfish) - has a small genome with little junk DNA
- The African clawed frog (*Xenopus laevis*) - used in developmental biology because of its large embryos and high tolerance for physical and pharmacological manipulation
- Zebrafish (*Danio rerio*, a freshwater fish) - has a nearly transparent body during early development, which provides unique visual access to the animal's internal anatomy. Zebrafish are used to study development, toxicology and toxicopathology, specific gene function and roles of signaling pathways.

Model Organisms Used for Specific Research Objectives

Sexual Selection and Sexual Conflict:

- *Callosobruchus maculatus*, the bruchid beetle
- *Chorthippus parallelus*, the meadow grasshopper
- Coelopidae - seaweed flies
- Diopsidae - stalk-eyed flies
- *Drosophila* spp. - fruit flies
- *Macrostomum lignano*, a sand flatworm
- *Gryllus bimaculatus*, the field cricket
- *Scathophaga stercoraria*, the yellow dung fly.

Hybrid Zones

* *Bombina bombina* and *variegata*
* *Podisma* spp. in the Alps
* *Caledia captiva* (Orthoptera) in eastern Australia.

Ecological Genomics

* *Daphnia pulex*, an environmental indicator model organism.

Table of Model Genetic Organisms

This table indicates the status of the genome sequencing project for each organism as well as whether the organism exhibits homologous recombination.

Organism	*Genome Sequenced*	*Homologous Recombination*
Prokaryote		
Escherichia coli	Yes	Yes
Eukaryote, unicellular		
Dictyostelium discoideum	Yes	Yes
Saccharomyces cerevisiae	Yes	Yes
Schizosaccharomyces pombe	Yes	Yes
Chlamydomonas reinhardtii	Yes	No
Tetrahymena thermophila	Yes	Yes
Eukaryote, multicellular		
Caenorhabditis elegans	Yes	Difficult
Drosophila melanogaster	Yes	Difficult
Arabidopsis thaliana	Yes	No
Physcomitrella patens	Yes	Yes
Vertebrate		
Danio rerio	Yes	Yes
Mus musculus	Yes	Yes
Xenopus laevis (Note: and X. tropicalis)	Yes	No
Homo sapiens (Note:not a model organism)	Yes	Yes

Animal Testing

Animal testing, also known as animal experimentation, animal research, and in vivo testing, is the use of non-human animals in experiments. Worldwide it is estimated that the number of vertebrate animals—from zebrafish to non-human primates—ranges from the tens of millions to more than 100 million used annually. Invertebrates, mice, rats, birds, fish, frogs, and animals not yet weaned are not included in the figures; one estimate of mice and rats used in the United States alone in 2001 was 80 million. Most animals are euthanised after being used in an experiment. Sources of laboratory animals vary between countries and species; most animals are purpose-bred, while others are caught in the wild or supplied by dealers who obtain them from auctions and pounds.

The research is conducted inside universities, medical schools, pharmaceutical companies, farms, defence establishments, and commercial facilities that provide animal-testing services to industry. It includes pure research such as genetics, developmental biology, behavioural studies, as well as applied research such as biomedical research, xenotransplantation, drug testing and toxicology tests, including cosmetics testing. Animals are also used for education, breeding, and defence research. The practice is regulated to various degrees in different countries.

Supporters of the use of animals in experiments, such as the British Royal Society, argue that virtually every medical achievement in the 20th century relied on the use of animals in some way, with the Institute for Laboratory Animal Research of the U.S. National Academy of Sciences arguing that even sophisticated computers are unable to model interactions between molecules, cells, tissues, organs, organisms, and the environment, making animal research necessary in many areas. Animal rights, and some animal welfare, organisations—such as PETA and BUAV—question the legitimacy of it, arguing that it is cruel, poor scientific practice, poorly regulated, that medical progress is being held back by misleading animal models, that some of the tests are outdated, that it cannot reliably predict effects in humans, that the costs outweigh the benefits, or that animals have an intrinsic right not to be used for experimentation.

Definitions

The terms animal testing, animal experimentation, animal research, *in vivo* testing, and vivisection have similar denotations but different connotations. Literally, "vivisection" means the "cutting up" of a living animal, and historically referred only to experiments that involved the dissection of live animals. The term is occasionally used to refer pejoratively to any experiment using living animals; for example, the *Encyclopædia Britannica* defines "vivisection" as: "Operation on a living animal for experimental rather than healing purposes; more broadly, all experimentation on live animals", although dictionaries point out that the broader definition is "used only by people who are opposed to such work". The word has a negative connotation, implying torture, suffering, and death. The word "vivisection" is preferred by those opposed to this research, whereas scientists typically use the term "animal experimentation".

History

The earliest references to animal testing are found in the writings of the Greeks in the 2nd and 4th centuries BCE. Aristotle (Áñéóôïô Ýëçò) (384–322 BCE) and Erasistratus (304–258 BCE) were among the first to perform experiments on living animals. Galen, a physician in 2nd-century Rome, dissected pigs and goats, and is known as the "father of vivisection." Avenzoar, an Arabic physician in 12th-century Moorish Spain who also practiced dissection, introduced animal testing as an experimental method of testing surgical procedures before applying them to human patients.

Animals have been used repeatedly through the history of biomedical research. The founders, in 1831, of the Dublin Zoo—the fourth oldest zoo in Europe, after Vienna, Paris, and London—were members of the medical profession, interested in studying the animals both while they were alive and when they were dead. In the 1880s, Louis Pasteur convincingly demonstrated the germ theory of medicine by inducing anthrax in sheep. In the 1890s, Ivan Pavlov famously used dogs to describe classical conditioning. Insulin was first isolated from dogs in 1922, and revolutionised the treatment of diabetes. On November 3, 1957, a Russian dog, Laika, became the first of many animals to orbit the earth. In the 1970s, antibiotic treatments and vaccines for leprosy were developed using armadillos, then given to humans. The ability of humans to change the genetics of animals took a large step forwards in 1974 when Rudolf Jaenisch was able to produce the first transgenic mammal, by integrating DNA from the SV40 virus into the genome of mice. This genetic research progressed rapidly and, in 1996, Dolly the sheep was born, the first mammal to be cloned from an adult cell.

Toxicology testing became important in the 20th century. In the 19th century, laws regulating drugs were more relaxed. For example, in the U.S., the government could only ban a drug after a company had been prosecuted for selling products that harmed customers. However, in response to the Elixir Sulfanilamide disaster of 1937 in which the eponymous drug killed more than 100 users, the U.S.

congress passed laws that required safety testing of drugs on animals before they could be marketed. Other countries enacted similar legislation. In the 1960s, in reaction to the Thalidomide tragedy, further laws were passed requiring safety testing on pregnant animals before a drug can be sold.

Historical Debate

As the experimentation on animals increased, especially the practice of vivisection, so did criticism and controversy. In 1655, the advocate of Galenic physiology Edmund O'Meara said that "the miserable torture of vivisection places the body in an unnatural state." O'Meara and others argued that animal physiology could be affected by pain during vivisection, rendering results unreliable. There were also objections on an ethical basis, contending that the benefit to humans did not justify the harm to animals. Early objections to animal testing also came from another angle — many people believed that animals were inferior to humans and so different that results from animals could not be applied to humans.

On the other side of the debate, those in favour of animal testing held that experiments on animals were necessary to advance medical and biological knowledge. Claude Bernard, known as the "prince of vivisectors" and the father of physiology—whose wife, Marie Françoise Martin, founded the first anti-vivisection society in France in 1883— famously wrote in 1865 that "the science of life is a superb and dazzlingly lighted hall which may be reached only by passing through a long and ghastly kitchen". Arguing that "experiments on animals ... are entirely conclusive for the toxicology and hygiene of man...the effects of these substances are the same on man as on animals, save for differences in degree," Bernard established animal experimentation as part of the standard scientific method.

In 1896, the physiologist and physician Dr. Walter B. Cannon said "The antivivisectionists are the second of the two types Theodore Roosevelt described when he said, 'Common sense without conscience may lead to crime, but conscience without common sense may lead to folly, which is the handmaiden of crime.' " These divisions between pro- and anti- animal testing groups first came to public attention during the brown dog affair in the early 1900s, when hundreds of medical students clashed with anti-vivisectionists and police over a memorial to a vivisected dog.

In 1822, the first animal protection law was enacted in the British parliament, followed by the Cruelty to Animals Act (1876), the first law specifically aimed at regulating animal testing. The legislation was promoted by Charles Darwin, who wrote to Ray Lankester in March 1871: "You ask about my opinion on vivisection. I quite agree that it is justifiable for real investigations on physiology; but not for mere damnable and detestable curiosity. It is a subject which makes me sick with horror, so I will not say another word about it, else I shall not sleep to-night." Opposition to the use of animals in medical research first arose in the United States during the 1860s, when Henry Bergh founded the American Society for the Prevention of Cruelty to Animals (ASPCA), with America's first specifically anti-vivisection organisation being the American AntiVivisection Society (AAVS), founded in 1883. Antivivisectionists of the era generally believed the spread of mercy was the great cause of civilization, and vivisection was cruel. However, in the USA the antivivisectionists' efforts were defeated in every legislature, overwhelmed by the superior organisation and influence of the medical community. Overall, this movement had little legislative success until the passing of the Laboratory Animal Welfare Act, in 1966.

Care and use of Animals

Regulations: The regulations that apply to animals in laboratories vary across species. In the U.S., under the provisions of the Animal Welfare Act and the *Guide for the Care and Use of Laboratory Animals* (the *Guide*), published by the National Academy of Sciences, any procedure can be performed on an animal if it can be successfully argued that it is scientifically justified. In general, researchers are required to consult with the institution's veterinarian and its Institutional Animal Care and Use Committee (IACUC), which every research facility is obliged to maintain.

The IACUC must ensure that alternatives, including non-animal alternatives, have been considered, that the experiments are not unnecessarily duplicative, and that pain relief is given unless it would interfere with the study. Larry Carbone, a laboratory animal veterinarian, writes that, in his experience, IACUCs take their work very seriously regardless of the species involved, though the use of non-human primates always raises what he calls a "red flag of special concern." A study published in *Science* magazine in July 2001 confirmed the low reliability of IACUC reviews of animal experiments. Funded by the National Science Foundation, the three-year study found that animal-use committees that do not know the specifics of the university and personnel do not make the same approval decisions as those made by animal-use committees that do know the university and personnel. Specifically, blinded committees more often ask for more information rather than approving studies.

The IACUCs regulate all vertebrates in testing at institutions receiving federal funds in the USA. Although the provisions of the Animal Welfare Act do not include purpose-bred rodents and birds, these species are equally regulated under Public Health Service policies that govern the IACUCs. Animal Welfare Act regulations are enforced by the USDA, whereas Public Health Service regulations are enforced by OLAW and in many cases by AAALAC.

Numbers

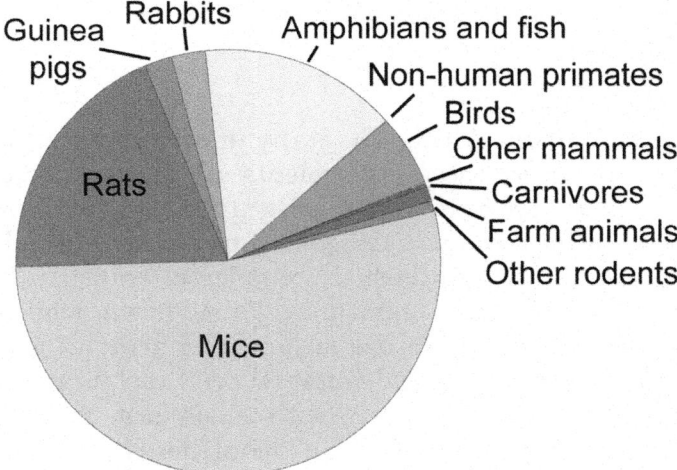

Figure 1: Types of vertebrates used in animal testing in Europe in 2005: a total of 12.1 million animals were used.

Accurate global figures for animal testing are difficult to obtain. The British Union for the Abolition of Vivisection (BUAV) estimates that 100 million vertebrates are experimented on around the world every year, 10–11 million of them in the European Union. The Nuffield Council on Bioethics reports that global annual estimates range from 50 to 100 million animals. None of the figures include invertebrates such as shrimp and fruit flies. Animals bred for research then killed as surplus, animals used for breeding purposes, and animals not yet weaned are also not included in the figures.

According to the U.S. Department of Agriculture (USDA), the total number of animals used in that country in 2005 was almost 1.2 million, but this does not include rats and mice, which make up about

90% of research animals. In 1995, researchers at Tufts University Centre for Animals and Public Policy estimated that 14–21 million animals were used in American laboratories in 1992, a reduction from a high of 50 million used in 1970. In 1986, the U.S. Congress Office of Technology Assessment reported that estimates of the animals used in the U.S. range from 10 million to upwards of 100 million each year, and that their own best estimate was at least 17 million to 22 million.

In the UK, Home Office figures show that 3.2 million procedures were carried out in 2007, a rise of 189,500 since the previous year. Four thousand procedures used non-human primates, down 240 from 2006. A "procedure" refers to an experiment that might last minutes, several months, or years. Most animals are used in only one procedure: animals either die because of the experiment or are euthanised afterwards.

Species

Invertebrates: Although many more invertebrates than vertebrates are used, these experiments are largely unregulated by law. The most used invertebrate species are *Drosophila melanogaster*, a fruit fly, and *Caenorhabditis elegans*, a nematode worm. In the case of *C. elegans*, the worm's body is completely transparent and the precise lineage of all the organism's cells is known, while studies in the fly *D. melanogaster* can use an amazing array of genetic tools. These animals offer great advantages over vertebrates, including their short life cycle and the ease with which large numbers may be studied, with thousands of flies or nematodes fitting into a single room. However, the lack of an adaptive immune system and their simple organs prevent worms from being used in medical research such as vaccine development. Similarly, flies are not widely used in applied medical research, as their immune system differs greatly from that of humans, and diseases in insects can be very different from diseases in vertebrates.

Vertebrates: In the U.S., the numbers of rats and mice used is estimated at 20 million a year. Other rodents commonly used are guinea pigs, hamsters, and gerbils. Mice are the most commonly used vertebrate species because of their size, low cost, ease of handling, and fast reproduction rate. Mice are widely considered to be the best model of inherited human disease and share 99% of their genes with humans. With the advent of genetic engineering technology, genetically modified mice can be generated to order and can provide models for a range of human diseases. Rats are also widely used for physiology, toxicology and cancer research, but genetic manipulation is much harder in rats than in mice, which limits the use of these rodents in basic science.

Nearly 200,000 fish and 20,000 amphibians were used in the UK in 2004. The main species used is the zebrafish, *Danio rerio*, which are translucent during their embryonic stage, and the African clawed frog, *Xenopus laevis*. Over 20,000 rabbits were used for animal testing in the UK in 2004. Albino rabbits are used in eye irritancy tests because rabbits have less tear flow than other animals, and the lack of eye pigment in albinos make the effects easier to visualize. Rabbits are also frequently used for the production of polyclonal antibodies.

Cats and Dogs

Cats are most commonly used in neurological research. Over 25,500 cats were used in the U.S. in 2000, around half of whom were used in experiments which, according to the American Anti-Vivisection Society, had the potential to cause "pain and/or distress".

Dogs are widely used in biomedical research, testing, and education — particularly beagles, because they are gentle and easy to handle. They are commonly used as models for human diseases in cardiology, endocrinology, and bone and joint studies, research that tends to be highly invasive, according to the Humane Society of the United States. The U.S. Department of Agriculture's Animal Welfare Report for 2005 shows that 66,000 dogs were used in USDA-registered facilities in that year. In the U.S., some of the dogs are purpose-bred, while most are supplied by so-called Class B dealers licensed by the USDA to buy animals from auctions, shelters, newspaper ads, and who are sometimes accused of stealing pets.

Non-human Primates

Non-human primates (NHPs) are used in toxicology tests, studies of AIDS and hepatitis, studies of neurology, behaviour and cognition, reproduction, genetics, and xenotransplantation. They are caught in the wild or purpose-bred. In the U.S. and China, most primates are domestically purpose-bred, whereas in Europe the majority are imported purpose-bred. Rhesus monkeys, cynomolgus monkeys, squirrel monkeys, and owl monkeys are imported; around 12,000 to 15,000 monkeys are imported into the U.S. annually. In total, around 70,000 NHPs are used each year in the United States and European Union. Most of the NHPs used are macaques; but marmosets, spider monkeys, and squirrel monkeys are also used, and baboons and chimpanzees are used in the U.S; in 2006 there were 1133 chimpanzees in U.S. primate centres. The first transgenic primate was produced in 2001, with the development of a method that could introduce new genes into a rhesus macaque. This transgenic technology is now being applied in the search for a treatment for the genetic disorder Huntington's disease. Notable studies on non-human primates have been part of the polio vaccine development, and development of Deep Brain Stimulation, and their current heaviest non-toxicological use occurs in the monkey AIDS model, SIV. In 2008 a proposal to ban all primates experiments in the EU has sparked a vigorous debate.

Sources

Animals used by laboratories are largely supplied by specialist dealers. Sources differ for vertebrate and invertebrate animals. Most laboratories breed and raise flies and worms themselves, using strains and mutants supplied from a few main stock centres. For vertebrates, sources include breeders who supply purpose-bred animals; businesses that trade in wild animals; and dealers who supply animals sourced from pounds, auctions, and newspaper ads. Animal shelters also supply the laboratories directly. Large centres also exist to distribute strains of genetically-modified animals; the National Institutes of Health *Knockout Mouse Project*, for example, aims to provide knockout mice for every gene in the mouse genome.

In the U.S., Class A breeders are licensed by the U.S. Department of Agriculture (USDA) to sell animals for research purposes, while Class B dealers are licensed to buy animals from "random sources" such as auctions, pound seizure, and newspaper ads. Some Class B dealers have been accused of kidnapping pets and illegally trapping strays, a practice known as *bunching*. It was in part out of public concern over the sale of pets to research facilities that the 1966 Laboratory Animal Welfare Act was ushered in — the Senate Committee on Commerce reported in 1966 that stolen pets had been retrieved from Veterans Administration facilities, the Mayo Institute, the University of Pennsylvania, Stanford University, and Harvard and Yale Medical Schools. The USDA recovered at least a dozen stolen pets during a raid on a Class B dealer in Arkansas in 2003.

Four states in the U.S. — Minnesota, Utah, Oklahoma, and Iowa — require their shelters to provide animals to research facilities. Fourteen states explicitly prohibit the practice, while the remainder either allow it or have no relevant legislation.

In the European Union, animal sources are governed by *Council Directive 86/609/EEC*, which requires lab animals to be specially bred, unless the animal has been lawfully imported and is not a wild animal or a stray. The latter requirement may also be exempted by special arrangement. In the UK, most animals used in experiments are bred for the purpose under the 1988 Animal Protection Act, but wild-caught primates may be used if exceptional and specific justification can be established. The United States also allows the use of wild-caught primates; between 1995 and 1999, 1,580 wild baboons were imported into the U.S. Over half the primates imported between 1995 and 2000 were handled by Charles River Laboratories, Inc., or by Covance, which is the single largest importer of primates into the U.S.

Pain and Suffering

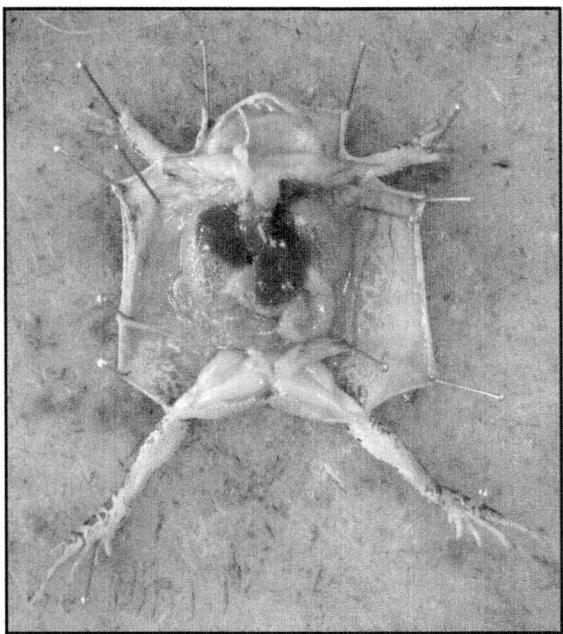

Figure 2: Prior to dissection for educational purposes, chloroform was administered to this common sand frog to induce anesthesia and death.

The extent to which animal testing causes pain and suffering, and the capacity of animals to experience and comprehend them, is the subject of much debate.

According to the U.S. Department of Agriculture, in 2006 about 670,000 animals (57%) (not including rats, mice, birds, or invertebrates) were used in procedures that did not include more than momentary pain or distress. About 420,000 (36%) were used in procedures in which pain or distress was relieved by anesthesia, while 84,000 (7%) were used in studies that would cause pain or distress that would not be relieved.

In the UK, research projects are classified as mild, moderate, and substantial in terms of the suffering the researchers conducting the study say they may cause; a fourth category of "unclassified" means the animal was anesthetised and killed without recovering consciousness, according to the researchers. In December 2001, 1,296 (39%) of project licenses in force were classified as mild, 1,811 (55%) as moderate, 63 (2%) as substantial, and 139 (4%) as unclassified. There have, however, been suggestions of systemic underestimation of procedure severity.

The idea that animals might not feel pain as human beings feel it traces back to the 17th-century French philosopher, René Descartes, who argued that animals do not experience pain and suffering because they lack consciousness. Bernard Rollin of Colorado State University, the principal author of two U.S. federal laws regulating pain relief for animals, writes that researchers remained unsure into the 1980s as to whether animals experience pain, and that veterinarians trained in the U.S. before 1989 were simply taught to ignore animal pain. In his interactions with scientists and other veterinarians, he was regularly asked to "prove" that animals are conscious, and to provide "scientifically acceptable" grounds for claiming that they feel pain. Carbone writes that the view that animals feel pain differently is now a minority view. Academic reviews of the topic are more equivocal, noting that although the argument that animals have at least simple conscious thoughts and feelings has strong support, some critics continue to question how reliably animal mental states can be determined. The ability of invertebrate species of animals, such as insects, to feel pain and suffering is also unclear.

The defining text on animal welfare regulation, "Guide for the Care and Use of Laboratory Animals" defines the parameters that govern animal testing in the USA. It states "The ability to experience and respond to pain is widespread in the animal kingdom...Pain is a stressor and, if not relieved, can lead to unacceptable levels of stress and distress in animals." The Guide states that the ability to recognise the symptoms of pain in different species is vital in efficiently applying pain relief and that it is essential for the people caring for and using animals to be entirely familiar with these symptoms. On the subject of analgesics used to relieve pain, the Guide states "The selection of the most appropriate analgesic or anesthetic should reflect professional judgment as to which best meets clinical and humane requirements without compromising the scientific aspects of the research protocol". Accordingly, all issues of animal pain and distress, and their potential treatment with analgesia and anesthesia, are required regulatory issues in receiving animal protocol approval.

Euthanasia

There is general agreement that animal life should not be taken wantonly, and regulations require that scientists use as few animals as possible. However, while policy makers consider suffering to be the central issue and see animal euthanasia as a way to reduce suffering, others, such as the RSPCA, argue that the lives of laboratory animals have intrinsic value. Regulations focus on whether particular methods cause pain and suffering, not whether their death is undesirable in itself. The animals are euthanised at the end of studies for sample collection or post-mortem examination; during studies if their pain or suffering falls into certain categories regarded as unacceptable, such as depression, infection that is unresponsive to treatment, or the failure of large animals to eat for five days; or when they are unsuitable for breeding or unwanted for some other reason.

Methods of euthanising laboratory animals are chosen to induce rapid unconsciousness and death without pain or distress. The methods that are preferred are those published by councils of veterinarians. The animal can be made to inhale a gas, such as carbon monoxide and carbon dioxide, by being placed in a chamber, or by use of a face mask, with or without prior sedation or anesthesia. Sedatives or anesthetics such as barbiturates can be given intravenously, or inhalant anesthetics may be used. Amphibians and fish may be immersed in water containing an anesthetic such as tricaine. Physical methods are also used, with or without sedation or anesthesia depending on the method. Recommended methods include decapitation (beheading) for small rodents or rabbits. Cervical dislocation (breaking the neck or spine) may be used for birds, mice, and immature rats and rabbits. Maceration (grinding into small pieces) is used on 1 day old chicks. High-intensity microwave irradiation of the brain can preserve brain tissue and induce death in less than 1 second, but this is currently only used on rodents. Captive bolts may be used, typically on dogs, ruminants, horses, pigs and rabbits. It causes death by a concussion to the brain. Gunshot may be used, but only in cases where a penetrating captive bolt may not be used. Some physical methods are only acceptable after the animal is unconscious. Electrocution may be used for cattle, sheep, swine, foxes, and mink after the animals are unconscious, often by a prior electrical stun. Pithing (inserting a tool into the base of the brain) is usable on animals already unconscious. Slow or rapid freezing, or inducing air embolism are acceptable only with prior anesthesia to induce unconsciousness.

Research Classification

Pure research: Basic or pure research investigates how organisms behave, develop, and function. Those opposed to animal testing object that pure research may have little or no practical purpose, but researchers argue that it may produce unforeseen benefits, rendering the distinction between pure and applied research—research that has a specific practical aim—unclear. Pure research uses larger numbers and a greater variety of animals than applied research. Fruit flies, nematode worms, mice and rats together account for the vast majority, though small numbers of other species are used, ranging from sea slugs through to armadillos. Examples of the types of animals and experiments used in basic research include:

- Studies on *embryogenesis* and *developmental biology*. Mutants are created by adding transposons into their genomes, or specific genes are deleted by gene targeting. By studying the changes in development these changes produce, scientists aim to understand both how organisms normally develop, and what can go wrong in this process. These studies are particularly powerful since the basic controls of development, such as the homeobox genes, have similar functions in organisms as diverse as fruit flies and man.

- Experiments into *behaviour*, to understand how organisms detect and interact with each other and their environment, in which fruit flies, worms, mice, and rats are all widely used. Studies of brain function, such as memory and social behaviour, often use rats and birds. For some species, behavioural research is combined with enrichment strategies for animals in captivity because it allows them to engage in a wider range of activities.

- Breeding experiments to study *evolution* and *genetics*. Laboratory mice, flies, fish, and worms are inbred through many generations to create strains with defined characteristics. These provide animals of a known genetic background, an important tool for genetic analyses. Larger mammals are rarely bred specifically for such studies due to their slow rate of reproduction, though some scientists take advantage of inbred domesticated animals, such as dog or cattle breeds, for comparative purposes. Scientists studying how animals evolve use many animal species to see how variations in where and how an organism lives (their niche) produce adaptations in their physiology and morphology. As an example, sticklebacks are now being used to study how many and which types of mutations are selected to produce adaptations in animals' morphology during the evolution of new species.

Applied Research

Applied research aims to solve specific and practical problems. Compared to pure research, which is largely academic in origin, applied research is usually carried out in the pharmaceutical industry, or by universities in commercial partnerships. These may involve the use of animal models of diseases or conditions, which are often discovered or generated by pure research programmes. In turn, such applied studies may be an early stage in the drug discovery process. Examples include:

- Genetic modification of animals to study disease. Transgenic animals have specific genes inserted, modified or removed, to mimic specific conditions such as single gene disorders, such as Huntington's disease. Other models mimic complex, multifactorial diseases with genetic components, such as diabetes, or even transgenic mice that carry the same mutations that occur during the development of cancer. These models allow investigations on how and why the disease develops, as well as providing ways to develop and test new treatments. The vast majority of these transgenic models of human disease are lines of mice, the mammalian species in which genetic modification is most efficient. Smaller numbers of other animals are also used, including rats, pigs, sheep, fish, birds, and amphibians.

- Studies on models of naturally occurring disease and condition. Certain domestic and wild animals have a natural propensity or predisposition for certain conditions that are also found in humans. Cats are used as a model to develop immunodeficiency virus vaccines and to study leukemia because their natural predisposition to FIV and Feline leukemia virus. Certain breeds of dog suffer from narcolepsy making them the major model used to study the human condition. Armadillos and humans are among only a few animal species that naturally suffer from leprosy; as the bacteria responsible for this disease cannot yet be grown in culture, armadillos are the primary source of bacilli used in leprosy vaccines.

- Studies on induced animal models of human diseases. Here, an animal is treated so that it develops pathology and symptoms that resemble a human disease. Examples include restricting blood flow to the brain to induce stroke, or giving neurotoxins that cause damage similar to that seen in Parkinson's disease. Such studies can be difficult to interpret, and it is argued that they are not always comparable to human diseases. For example, although such models are now widely used to study Parkinson's disease, the British anti-vivisection interest group BUAV argues that these models only superficially resemble the disease symptoms, without the same time course or cellular pathology. In contrast, scientists assessing the usefulness of animal models of Parkinson's disease, as well as the medical research charity *The Parkinson's Appeal*, state that these models were invaluable and that they led to improved surgical treatments such as pallidotomy, new drug treatments such as levodopa, and later deep brain stimulation.

Xenotransplantation

Xenotransplantation research involves transplanting tissues or organs from one species to another, as a way to overcome the shortage of human organs for use in organ transplants. Current research involves using primates as the recipients of organs from pigs that have been genetically-modified to reduce the primates' immune response against the pig tissue. Although transplant rejection remains a problem, recent clinical trials that involved implanting pig insulin-secreting cells into diabetics did reduce these people's need for insulin.

Documents released to the news media by the animal rights organisation Uncaged Campaigns showed that, between 1994 and 2000, wild baboons imported to the UK from Africa by Imutran Ltd, a subsidiary of Novartis Pharma AG, in conjunction with Cambridge University and Huntingdon Life Sciences, to be used in experiments that involved grafting pig tissues, suffered serious and sometimes fatal injuries. A scandal occurred when it was revealed that the company had communicated with the British government in an attempt to avoid regulation.

Toxicology Testing

Toxicology testing, also known as safety testing, is conducted by pharmaceutical companies testing drugs, or by contract animal testing facilities, such as Huntingdon Life Sciences, on behalf of a wide variety of customers. According to 2005 EU figures, around one million animals are used every year in Europe in toxicology tests; which are about 10% of all procedures. According to *Nature*, 5,000 animals are used for each chemical being tested, with 12,000 needed to test pesticides. The tests are conducted without anesthesia, because interactions between drugs can affect how animals detoxify chemicals, and may interfere with the results.

Toxicology tests are used to examine finished products such as pesticides, medications, food additives, packing materials, and air freshener, or their chemical ingredients. Most tests involve testing ingredients rather than finished products, but according to BUAV, manufacturers believe these tests overestimate the toxic effects of substances; they therefore repeat the tests using their finished products to obtain a less toxic label.

The substances are applied to the skin or dripped into the eyes; injected intravenously, intramuscularly, or subcutaneously; inhaled either by placing a mask over the animals and restraining them, or by placing them in an inhalation chamber; or administered orally, through a tube into the stomach, or simply in the animal's food. Doses may be given once, repeated regularly for many months, or for the lifespan of the animal.

There are several different types of acute toxicity tests. The LD_{50} ("Lethal Dose 50%") test is used to evaluate the toxicity of a substance by determining the dose required to kill 50% of the test animal population. This test was removed from OECD international guidelines in 2002, replaced by methods such as the fixed dose procedure, which use fewer animals and cause less suffering. *Nature* writes that, as of 2005, "the LD50 acute toxicity test ... still accounts for one-third of all animal [toxicity] tests worldwide." Irritancy can be measured using the Draize test, where a test substance is applied to an animal's eyes or skin, usually an albino rabbit. For Draize eye testing, the test involves observing the effects of the substance at intervals and grading any damage or irritation, but the test should be halted and the animal killed if it shows "continuing signs of severe pain or distress". The Humane Society of the United States writes that the procedure can cause redness, ulceration, hemorrhaging, cloudiness, or even blindness. This test has also been criticized by scientists for being cruel and inaccurate, subjective, over-sensitive, and failing to reflect human exposures in the real world. Although no accepted *in vitro* alternatives exist, a modified form of the Draize test called the *low volume eye test* may reduce suffering and provide more realistic results and this was adopted as the new standard in September 2009. However, the Draize test will still be used for substances that are not severe irritants.

The most stringent tests are reserved for drugs and foodstuffs. For these, a number of tests are performed, lasting less than a month (acute), one to three months (subchronic), and more than three months (chronic) to test general toxicity (damage to organs), eye and skin irritancy, mutagenicity, carcinogenicity, teratogenicity, and reproductive problems. The cost of the full complement of tests is several million dollars per substance and it may take three or four years to complete.

These toxicity tests provide, in the words of a 2006 United States National Academy of Sciences report, "critical information for assessing hazard and risk potential". *Nature* reported that most animal tests either over- or underestimate risk, or do not reflect toxicity in humans particularly well, with false positive results being a particular problem. This variability stems from using the effects of high doses of chemicals in small numbers of laboratory animals to try to predict the effects of low doses in large numbers of humans. Although relationships do exist, opinion is divided on how to use data on one species to predict the exact level of risk in another.

Cosmetics Testing

Cosmetics testing on animals is particularly controversial. Such tests, which are still conducted in the U.S., involve general toxicity, eye and skin irritancy, phototoxicity (toxicity triggered by ultraviolet light) and mutagenicity.

Cosmetics testing is banned in the Netherlands, Belgium, and the UK, and in 2002, after 13 years of discussion, the European Union (EU) agreed to phase in a near-total ban on the sale of animal-tested cosmetics throughout the EU from 2009, and to ban all cosmetics-related animal testing. France, which is home to the world's largest cosmetics company, L'Oreal, has protested the proposed ban by lodging a case at the European Court of Justice in Luxembourg, asking that the ban be quashed. The ban is also opposed by the European Federation for Cosmetics Ingredients, which represents 70 companies in Switzerland, Belgium, France, Germany and Italy.

Drug Testing

Before the early 20th century, laws regulating drugs were lax. Currently, all new pharmaceuticals undergo rigorous animal testing before being licensed for human use. Tests on pharmaceutical products involve:

- *metabolic tests*, investigating pharmacokinetics – how drugs are absorbed, metabolised and excreted by the body when introduced orally, intravenously, intraperitoneally, intramuscularly, or transdermally.

- *toxicology tests*, which gauge acute, subacute, and chronic toxicity. Acute toxicity is studied by using a rising dose until signs of toxicity become apparent. Current European legislation demands that "acute toxicity tests must be carried out in two or more mammalian species" covering "at least two different routes of administration". Subacute toxicity is where the drug is given to the animals for four to six weeks in doses below the level at which it causes rapid poisoning, in order to discover if any toxic drug metabolites build up over time. Testing for chronic toxicity can last up to two years and, in the European Union, is required to involve two species of mammals, one of which must be non-rodent.

- *efficacy studies*, which test whether experimental drugs work by inducing the appropriate illness in animals. The drug is then administered in a double-blind controlled trial, which allows researchers to determine the effect of the drug and the dose-response curve.

- Specific tests on *reproductive function, embryonic toxicity*, or *carcinogenic potential* can all be required by law, depending on the result of other studies and the type of drug being tested.

Education, Breeding, and Defence

Animals are also used for education and training; are bred for use in laboratories; and are used by the military to develop weapons, vaccines, battlefield surgical techniques, and defensive clothing. For example, in 2008 the United States Defence Advanced Research Projects Agency used live pigs to study the effects of improvised explosive device explosions on internal organs, especially the brain.

There are efforts in many countries to find alternatives to using animals in education. Horst Spielmann, German director of the Central Office for Collecting and Assessing Alternatives to Animal

Experimentation, while describing Germany's progress in this area, told German broadcaster ARD in 2005: "Using animals in teaching curricula is already superfluous. In many countries, one can become a doctor, vet or biologist without ever having performed an experiment on an animal."

Ethics

Viewpoints: The ethical questions raised by performing experiments on animals are subject to much debate, and viewpoints have shifted significantly over the 20th century. There remain disagreements about which procedures are useful for which purposes, as well as disagreements over which ethical principles apply to which species. The dominant ethical position worldwide is that achievement of scientific and medical goals using animal testing is desirable, so long as animal suffering and use is minimised. The British government has additionally required that the cost to animals in an experiment be weighed against the gain in knowledge. Some medical schools and agencies in China, Japan, and South Korea have built cenotaphs for killed animals. In Japan there are also annual memorial services (*Ireisai* pa—my) for animals sacrificed at medical school.

A wide range of minority viewpoints exist. The view that animals have moral rights (animal rights) is a philosophical position proposed by Tom Regan, among others, who argues that animals are beings with beliefs and desires, and as such are the "subjects of a life" with moral value and therefore moral rights. Regan still sees ethical differences between killing human and non-human animals, and argues that to save the former it is permissible to kill the latter. Likewise, a "moral dilemma" view suggests that avoiding potential benefit to humans is unacceptable on similar grounds, and holds the issue to be a dilemma in balancing such harm to humans to the harm done to animals in research. In contrast, an abolitionist view in animal rights holds that there is no moral justification for any harmful research on animals that is not to the benefit of the individual animal. Bernard Rollin argues that benefits to human beings cannot outweigh animal suffering, and that human beings have no moral right to use an animal in ways that do not benefit that individual. Another prominent position is that of philosopher Peter Singer, who argues that there are no grounds to include a being's species in considerations of whether their suffering is important in utilitarian moral considerations.

Although these arguments have not been widely accepted, governments such as the Netherlands and New Zealand have responded to the concerns by outlawing invasive experiments on certain classes of non-human primates, particularly the great apes.

Prominent Cases

Various specific cases of animal testing have drawn attention, including both instances of beneficial scientific research, and instances of alleged ethical violations by those performing the tests.

Muscle Physiology

The fundamental properties of muscle physiology were determined with on work done using frog muscles (including the force generating mechanism of all muscle, the length-tension relationship, and the force-velocity curve), and frogs are still the preferred model organism due to the long survival of muscles in vitro and the possibility of isolating intact single-fibre preparations (not possible in other organisms). Modern physical therapy and the understanding and treatment of muscular disorders is based on this work and subsequent work in mice (often engineered to express disease states such as muscular dystrophy).

University of California, Riverside

1985 was a pivotal year in the debate about animal research in the United States, with the enactment of amendments to the Animal Welfare Act. Britches, a macaque monkey, was born that year inside the University of California, Riverside, removed from his mother at birth, and left alone with his eyelids sewn shut, and a sonar sensor on his head, as part of an experiment to test sensory substitution devices for blind people. The Animal Liberation Front raided the laboratory on April 20, 1985, removing Britches and 466 other animals, and reportedly inflicting $700,000-worth of damage to equipment. A spokesman for the university said the allegations of mistreatment were false, and that the raid caused long-term damage to its research projects. The National Institutes of Health conducted an eight-month investigation and concluded that no corrective action was necessary.

Huntingdon Life Sciences

In 1997, People for the Ethical Treatment of Animals filmed staff inside Huntingdon Life Sciences (HLS) in the UK, Europe's largest animal-testing facility, hitting puppies, shouting at them, and simulating sex acts while taking blood samples. The company said the employees were dismissed. Two pleaded guilty to "cruelly terrifying dogs," and were given community service orders and ordered to pay £250 costs, the first lab technicians to have been prosecuted for animal cruelty in the UK. The broadcast of the video on Britain's Channel 4 Television in March 1997 triggered the formation of Stop Huntingdon Animal Cruelty (SHAC), an international leaderless resistance campaign to close HLS, which has been criticized for its sometimes violent tactics. In January 2009, several British SHAC activists were jailed for blackmailing companies linked to HLS.

Roslin Institute

In February 1997 a team at the Roslin Institute in Scotland announced the birth of Dolly the sheep, a ewe that had been cloned from tissue taken from another adult sheep. Dolly was produced through nuclear transfer to an unfertilized oocyte, and was the only lamb that survived from 277 attempts at this technique. Dolly appeared to be a normal sheep, living for six years and giving birth to several lambs, but was euthanised in 2003 after contracting a progressive lung disease. Although the production of Dolly was a scientific breakthrough, it was controversial, since it showed that not only could cloned animals be produced for use in farming, but also that it would now be, in principle, possible to clone a human being.

University of Cambridge

The British Union for the Abolition of Vivisection (BUAV) raised concerns about primate experiments at the University of Cambridge in 2002. In a series of court cases, the BUAV alleged that monkeys had undergone surgery to induce a stroke, and were left alone after the procedure for 15 hours overnight. Researchers had trained the monkeys to perform certain tasks before inflicting brain damage and re-testing them. The monkeys were only given food and water for two hours a day, to encourage them to perform the tasks. The judge hearing BUAV's application for a judicial review rejected the allegation that the Home Secretary had been negligent in granting the university a license. The British government's chief inspector of animals conducted a review of the facilities and experiments. It concluded the veterinary input at Cambridge was "exemplary"; the facility "seems adequately staffed"; and the animals afforded "appropriate standards of accommodation and care."

Columbia University

CNN reported in October 2003 that Catherine Dell'Orto, a veterinarian at Columbia University, had approached the university's Institute of Comparative Medicine about the treatment of baboons who were undergoing surgery as part of an experiment into stroke treatment. She said the baboons, who were in some cases having an eyeball removed, were left to suffer in their cages after the surgery. She alleged there was systemic maltreatment, poor record-keeping, and other violations of regulations, according to CNN. She presented her evidence in October 2002 and, dissatisfied with the response, contacted People for the Ethical Treatment of Animals two months later.

In March 2003, a lab technician shot video inside the lab, which according to *The New York Daily News* showed primates in cages without pain medication; the video included one baboon with a metal cylinder screwed into its head, according to the newspaper. Dell'Orto told the newspaper that primates were often not euthanised or given painkillers after surgery; she said other primates had torn their fingers off out of fear. The U.S. Department of Agriculture upheld Dell'Orto's complaint that there was shoddy record-keeping, and that 11 animals had been provided with "inadequate or questionable care." They found no evidence that the experiments violated federal guidelines or that there had been retaliation against Dell'Orto. CNN reported that Columbia responded by ordering better record-keeping, a review of the veterinary care program, and tighter criteria for euthanasia of laboratory animals.

Covance

In 2004, German journalist Friedrich Mülln shot undercover footage of staff in Covance, Münster, Europe's largest primate-testing centre, making monkeys dance in time to blaring pop music, handling them roughly, and screaming at them. The monkeys were kept isolated in small wire cages with little or no natural light, no environmental enrichment, and high noise levels from staff shouting and playing the radio (video). Primatologist Jane Goodall described the living conditions of the monkeys as horrendous. Another primatologist, Stephen Brend, told BUAV that using monkeys in such a stressed state is bad science, and trying to extrapolate useful data in such circumstances is what he called an untenable proposition. In 2004 and 2005, PETA shot footage inside the company in the United States. According to *The Washington Post*, PETA said an employee of the group filmed primates being choked, hit, and denied medical attention when badly injured. The U.S. Department of Agriculture fined Covance $8,720 for 16 citations, three of which involved lab monkeys; the other citations involved administrative issues and equipment.

Threats to Researchers

In 2006, a primate researcher at the University of California, Los Angeles (UCLA) shut down the experiments in his lab after threats from animal rights activists. The researcher had received a grant to use 30 macaque monkeys for vision experiments; each monkey was an esthetised for a single physiological experiment lasting up to 120 hours, and then euthanised. The researcher's name, phone number, and address were posted on the website of the Primate Freedom Project. Demonstrations were held in front of his home. A Molotov cocktail was placed on the porch of what was believed to be the home of another UCLA primate researcher; instead, it was accidentally left on the porch of an elderly woman unrelated to the university. The Animal Liberation Front claimed responsibility for the attack. As a result of the campaign, the researcher sent an email to the Primate Freedom Project stating "you win," and "please don't bother my family anymore." In another incident at UCLA in June 2007, the Animal Liberation Brigade placed a bomb under the car of a UCLA children's ophthalmologist who experiments on cats and rhesus monkeys; the bomb had a faulty fuse and did not detonate. UCLA is now refusing Freedom of Information Act requests for animal medical records.

These attacks, as well as similar incidents that caused the Southern Poverty Law Centre to declare in 2002 that the animal rights movement had "clearly taken a turn toward the more extreme," this prompted the US government to pass the Animal Enterprise Terrorism Act and the UK government to add the offence of "Intimidation of persons connected with animal research organisation" to the Serious Organised Crime and Police Act 2005. Such legislation, and the arrest and imprisonment of extremists may have decreased the incidence of attacks.

Alternatives to Animal Testing

Scientists and governments state that animal testing should cause as little suffering to animals as possible, and that animal tests should only be performed where necessary. The "three Rs" are guiding principles for the use of animals in research in most countries:

1. Replacement refers to the preferred use of non-animal methods over animal methods whenever it is possible to achieve the same scientific aim.

2. Reduction refers to methods that enable researchers to obtain comparable levels of information from fewer animals, or to obtain more information from the same number of animals.

3. Refinement refers to methods that alleviate or minimise potential pain, suffering or distress, and enhance animal welfare for the animals still used.

Although such principles have been welcomed as a step forwards by some animal welfare groups, they have also been criticized as both outdated by current research, and of little practical effect in improving animal welfare.

Animal Colouration

Animal colouration has been a topic of interest and research in biology for well over a century. According to Charles Darwin's 1859 theory of natural selection, features such as colouration evolved by providing individual animals with a reproductive advantage. For example, an individual with slightly better camouflage than others of the same species would on average leave more offspring.

There are at least six separate reasons why animal colouration may evolve:

- Camouflage, enabling an animal to remain hidden from view
- Warning, signalling to other animals not to attack
- Mimicry, taking advantage of another species' warning colouration
- Sexual selection, signalling to other members of the same species
- Simple physical protection, such as having pigments to protect against sunburn
- and purely incidental colouration, such as having red blood because, as it happens, haem (needed to carry oxygen) is red.

These reasons are briefly introduced below: they are covered in detail in separate articles.

Camouflage

One of the pioneers of research into animal colouration, Edward Bagnall Poulton classified the forms of protective colouration including camouflage in a way which is still helpful:

- Protective resemblance
  1. Special: the whole animal looks like some other object, for example when a caterpillar resembles a twig or a bird dropping. This is now called Mimesis
  2. General: the animal's texture blends with the background, for example when a moth's colour and pattern blend in with tree bark. This is now called crypsis

- Aggressive resemblance
    1. Special: a predator (or parasite) looks like something else, luring the prey to approach, for example when a flower mantis resembles a particular kind of flower, such as an orchid
    2. General: a predator or parasite blends in with the background, for example when a leopard is hard to see in long grass
- Adventitious protection: an animal uses materials such as twigs, sand, or pieces of shell to conceal its outline, for example when a Caddis Fly larva builds a decorated case, or when a crab decorates its back with seaweed, sponges and stones
- Variable protective resemblance: an animal such as a chameleon, flatfish, squid or octopus changes its skin pattern and colour using special chromatophore cells to resemble whatever background it is currently resting on (as well as for signalling).

The Theory of camouflage seeks to explain how resemblances such as those noted by Poulton are achieved. The main mechanisms, which apply equally in nature and in military applications are:

- Cryptic, blending into the background so as to become hard to see (this covers both special and general resemblance)
- Disruptive, using colour and pattern to break up the animal's outline (this relates mainly to general resemblance)
- Mimicry (in a narrow sense: other types of mimicry are described separately), resembling other objects of no special interest to the observer (this relates to special resemblance)
- Countershading, using graded colour to create the illusion of flatness (this relates mainly to general resemblance).

Countershading was first described by the American artist Abbott Handerson Thayer, a pioneer in the theory of animal colouration. Thayer observed that whereas a painter takes a flat canvas and uses coloured paint to create the illusion of solidity by painting in shadows, animals such as deer are often darkest on their backs, becoming lighter towards the belly, creating (as zoologist Hugh Cott observed) the illusion of flatness, and against a matching background, of invisibility. Thayer's observation "Animals are painted by Nature, darkest on those parts which tend to be most lighted by the sky's light, and *vice versa*" is called *Thayer's Law*.

Warning Colouration

Warning colouration (aposematism) is effectively the "opposite" of camouflage. Its function is to make the animal, for example a wasp or a coral snake, highly conspicuous to potential predators, so that it is noticed, remembered, and then avoided. As Peter Forbes observes, "Human warning signs employ the same colours - red, yellow, black, and white - that nature uses to advertise dangerous creatures." Warning colours work by being associated by potential predators with something that makes the warning-coloured animal unpleasant or dangerous. This can be achieved in several ways:

- distasteful, for example a Cinnabar moth caterpillar has bitter-tasting chemicals in its blood
- foul-smelling, for example the skunk can eject a liquid with a long-lasting and powerful odour
- poisonous, for example a wasp can deliver a painful sting, while a viper can deliver a fatal bite

Warning colouration can succeed either through inborn ("instinctual") behaviour on the part of potential predators, or through a learned avoidance. Either can lead to various forms of mimicry.

Mimicry

The existence of warning colouration (aposematism) makes it possible for mimicry to evolve, because it enables natural selection to drive slight, chance, resemblance to progressively more perfect mimicry. There are numerous possible mechanisms, of which by far the best known are:

- Batesian mimicry, the resemblance of edible to distasteful animals, most commonly insects such as butterflies; a familiar example is the resemblance of harmless hoverflies (which have no sting) to bees
- Müllerian mimicry, the mutual resemblances among distasteful animals, most commonly insects such as wasps and bees (hymenoptera)

Batesian mimicry was first described by pioneering naturalist Henry W. Bates. When an edible prey animal comes to resemble, even slightly, a distasteful animal (not necessarily closely related to it), natural selection favours those individuals that even very slightly better resemble the distasteful target. This is because even a small degree of protection reduces predation and increases the chance that an individual mimic will survive and reproduce. For example, many species of hoverfly are coloured black and yellow like bees, and are in consequence avoided by birds (and people).

Müllerian mimicry was first described by pioneering naturalist Fritz Müller. When a distasteful animal comes to resemble a more common distasteful animal, natural selection favours individuals that even very slightly better resemble the target. For example, many species of stinging wasp and bee are similarly coloured black and yellow. Müller's explanation of the mechanism for this was one of the first uses of mathematics in biology.

Müller's argument runs basically (using a simple example rather than equations) as follows:

1. Suppose there are, say, 100 wasps of rare species A and 1000 wasps of common species B in a place.

2. Suppose that wasps are eaten by young inexperienced birds, which quickly learn after one trial, by getting stung in the mouth, not to eat wasps again.

3. Suppose there are 10 young birds in the place.

4. If species A does not resemble B (to the birds), then each young bird must eat one A and one B to learn to avoid them. 10 out of 100 wasps of species A, and 10 out of 1000 wasps of species B perish in the training process. Note that in this example, A is 10 times rarer than B, and therefore suffers 10 times as heavily.

5. Now suppose that species A resembles B perfectly, so the young birds cannot distinguish them. Each young bird now needs only to eat one wasp - A or B, it doesn't matter - to learn to avoid both of them.

6. The advantage gained by species A of resembling species B is that where before 10 individuals of species A perished, now only about 1 perishes, as most of the wasps sampled by the young birds at random will belong to the commoner species B. Note that there is a large gain from the resemblance in the rare species, and a small gain (1 fewer individual out of 1000 perishes) in the common species: in fact, A's gain is 100 times as much as B's, comparing before and after.

7. Therefore there is a powerful selective pressure favouring progressively closer resemblance of species A to species B.

Darwin suggested an explanation of these differences in his theory of sexual selection (*The Descent of Man*, London, 1874): once the females begin to select males according to any particular characteristic, such as a long tail or a coloured crest, that characteristic will progressively be emphasized in the males. Eventually all the males will have the characteristics that the females are sexually selecting for strongly emphasized, as any male that does not will not reproduce. Note that this mechanism is so powerful that it is able to create features that are strongly disadvantageous to the males in other ways: for example, some male Birds of Paradise have wing or tail streamers that are so long that they may impede flight, while their brilliant colours may make the males more vulnerable to predators. In the extreme, it may be that sexual selection has driven species to extinction, as has been argued for the enormous horns of the male Irish Elk. Different forms of sexual selection are possible, including rivalry among males, and selection of females by males.

Physical Protection

Many animals have dark pigments such as melanin in their skin, eyes and fur to protect themselves against sunburn (damage to living tissues caused by ultraviolet light).

Incidental Colouration

Some animals are coloured purely incidentally because substances that they produce for other purposes happen to be pigments. For example, amphibians that live in caves may be largely colourless as colour has no function in that environment, but they may have red blood and show some red in their skin because the haem in their blood cells, needed to carry oxygen, happens to be red.

Blind Animals

Visual perception plays an important role in the animal kingdom, most notably for the identification of food sources, and avoidance of predators. For this reason, blindness in animals is a unique topic of study. In general, nocturnal or subterranean animals have less interest in the visual world, and depend on other sensory modalities. Visual capacity is a continuum, with humans falling somewhere in the centre.

Totally Blind Species

This list is incomplete; you can help by expanding it.

· Mole (animal)

The Star-nosed Mole can detect, catch and eat food faster than the human eye can follow (under 300 milliseconds).

Blind animals include the blind cave fish and cave crickets, the Texas salamander, blind flatworms, eyeless shrimp, eyeless fish, cave beetles, cave crayfish, and some bristletails, isopods and copepods.

Some animals live only in caves - they are called troglobites (meaning 'cave dwellers'). These animals are adapted to life in the dark.

Partial Blindness

Although the eyes of most bat species are small and poorly developed, leading to poor visual acuity, it is incorrect to assume that they are nearly blind. Vision is used as an aid in navigation especially at long distances, beyond the range of echolocation.

Infant Blindness

Blindness at birth serves to preserve the young who are dependent on their parents. (If they could see, they could wander off.) Rabbits are born with eyes and ears closed, totally helpless. Humans have very poor vision at birth as well.

Statements that certain species of mammals are "born blind" refers to them being born with their eyes closed and their eyelids fused together; the eyes open later. One example is the rabbit. In humans the eyelids are fused for a while before birth, but open again before the normal birth time, but very premature babies are sometimes born with their eyes fused shut, and opening later. Other animals such as the blind mole rat are truly blind and rely on other senses.

Colour Blindness

Humans and primates are unique as they possess trichromatic colour vision, and are able to discern between violet [short wave (SW)], green [medium wave (MW)], and yellow-green [long wave (LW)]. Mammals other than primates generally have less effective two-receptor colour perception systems, allowing only dichromatic colour vision; marine mammals have only a single cone type and are thus monochromats. Honey- and bumblebees have trichromatic colour vision, which is insensitive to red but sensitive in ultraviolet to a colour called *bee purple*.

Other animals, such as tropical fish and birds, have more complex colour vision systems than humans. There is evidence that ultraviolet light plays a part in colour perception in many branches of the animal kingdom, (especially insects) though there has not been enough evidence to show this. It has been suggested that it is likely that pigeons are pentachromats. *Papilio* butterflies apparently have tetrachromatic colour vision despite possessing six photoreceptor types. The most complex colour vision system in animal kingdom has been found in stomatopods with up to 12 different spectral receptor types which are thought to work as multiple dichromatic units.

Natural Selection

Darwin cites moles as an example of mammals that have organs that have become vestigial and are being phased out by natural selection:

> The eyes of moles and of some burrowing rodents are rudimentary in size, and in some cases are quite covered by skin and fur. This state of the eyes is probably due to gradual reduction from disuse, but aided perhaps by natural selection. In South America, a burrowing rodent, the tuco-tuco, or Ctenomys, is even more subterranean in its habits than the mole; and I was assured by a Spaniard, who had often caught them, that they were frequently blind. One which I kept alive was certainly in this condition, the cause, as appeared on dissection, having been inflammation of the nictitating membrane. As frequent inflammation of the eyes must be injurious to any animal, and as eyes are certainly not necessary to animals having subterranean habits, a reduction in their size, with the adhesion of the eyelids and growth of fur over them, might in such case be an advantage; and if so, natural selection would aid the effects of disuse.

Research

The blind forms of the Mexican tetra have proven popular subjects for scientists studying evolution: A recent study suggests that there are at least two distinct genetic lineages among the blind populations, arguing that these represent a case of convergent evolution.

One theory is that because of its dark habitat, the fish embryo saves energy it would normally use to develop eyes to develop other body parts, and this developmental choice would eventually dominate the population. This is called economical adaptation. However, studies have shown that blind cave fish embryos begin to grow eyes during development but then something actively stops this process and flesh grows over the partially grown eyes. Another theory is that some Mexican tetra randomly don't develop eyes (which is represented by broken genes in the fish's genome), and this lack of eyes spreads to the rest of the population despite having no advantage or disadvantage. This is called the unified neutral theory of biodiversity.

In one experiment studying eye development, University of Maryland scientists transplanted lenses from the eyes of sighted surface-form embryos into blind cave-form embryos, and vice versa. In the cave form, lens development begins within the first 24 hours of embryonic development, but quickly aborts, the lens cells dying; most of the rest of the eye structures never develop. Researchers found that the lens seemed to control the development of the rest of the eye, as the surface-form tetras which received cave-form lenses failed to develop eyes, while cave-form tetras which received surface-form lenses grew eyes with pupils, corneas, and irises. (It is not clear whether they possessed sight, however.)

The evolution of trichromatic colour vision in primates occurred as the ancestors of modern monkeys, apes, and humans switched to diurnal (daytime) activity and began consuming fruits and leaves from flowering plants.

Injury, Disease and Disability

Blindness often afflicts pets, especially glaucoma in old dogs.

In Fiction

The theme of blind animals has been a powerful one in literature. Peter Schaffer's Tony-Award winning play, Equus, tells the story of a boy who blinds six horses. Theodore Taylor's classic young adult novel, *The Trouble With Tuck,* is about a teenage girl, Helen, who trains her blind dog to follow and trust a seeing-eye dog. In non-fiction, a recent classic is Linda Kay Hardie's essay, "Lessons Learned from a Blind Cat," in *Cat Women: Female Writers on their Feline Friends.*

Reproduction

Reproduction (or procreation) is the biological process by which new "offspring" individual organisms are produced from their "parents". Reproduction is a fundamental feature of all known life; each individual organism exists as the result of reproduction. The known methods of reproduction are broadly grouped into two main types: sexual and asexual.

In asexual reproduction, an individual can reproduce without involvement with another individual of that species. The division of a bacterial cell into two daughter cells is an example of asexual reproduction. Asexual reproduction is not, however, limited to single-celled organisms. Most plants have the ability to reproduce a sexually and the ant species. Mycocepurus smithii is thought to reproduce entirely by asexual means.

Sexual reproduction typically requires the involvement of two individuals or gametes, one each from opposite type of sex.

Reproductive System

In biological terms sexual reproduction involves the union of gametes - the sperm and the ovum - produced by two parents. Each gamete is formed by meiosis. This means each contains only half the chromosomes of the body cells (haploid). Fertilization results in the joining of the male and female gametes to form a zygote which contains the full number of chromosomes (diploid). The zygote then starts to divide by mitosis to form a new animal with all its body cells containing chromosomes that are identical to those of the original zygote.

The offspring formed by sexual reproduction contain genes from both parents and show considerable variation. For example, kittens in a litter are all different although they (usually) have the same mother and father. In the wild this variation is important because it means that when the environment changes some individuals may be better adapted to survive than others. These survivors pass their "superior" genes on to their offspring. In this way the characteristics of a group of animals can gradually change over time to keep pace with the changing environment. This "survival of the fittest" or "natural selection" is the mechanism behind the theory of evolution.

Fertilization

In most fish and amphibia (frogs and toads) fertilization of the egg cells takes place outside the body. The female lays the eggs and then the male deposits his sperm on or at least near them.

In reptiles and birds, eggs are fertilized inside the body when the male deposits the sperm inside the egg duct of the female. The egg is then surrounded by a resistant shell, "laid" by the female and the embryo completes its development inside the egg.

In mammals the sperm are placed in the body of the female and the eggs are fertilized internally. They then develop to quite an advanced stage inside the body of the female. When they are born they are fed on milk excreted from the mammary glands and protected by their parents until they become independent.

The Male Reproductive System

The male reproductive system consists of a pair of testes that produce sperm (or spermatozoa), ducts that transport the sperm to the penis and glands that add secretions to the sperm to make semen. The various parts of the male reproductive system with a summary of their functions are shown as:

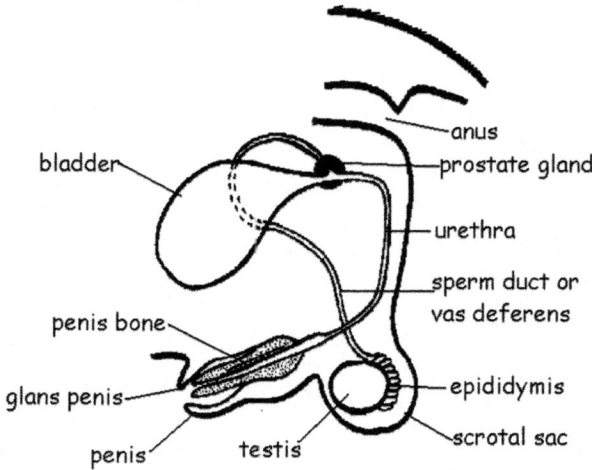

Figure 3: The reproductive organs of a male dog

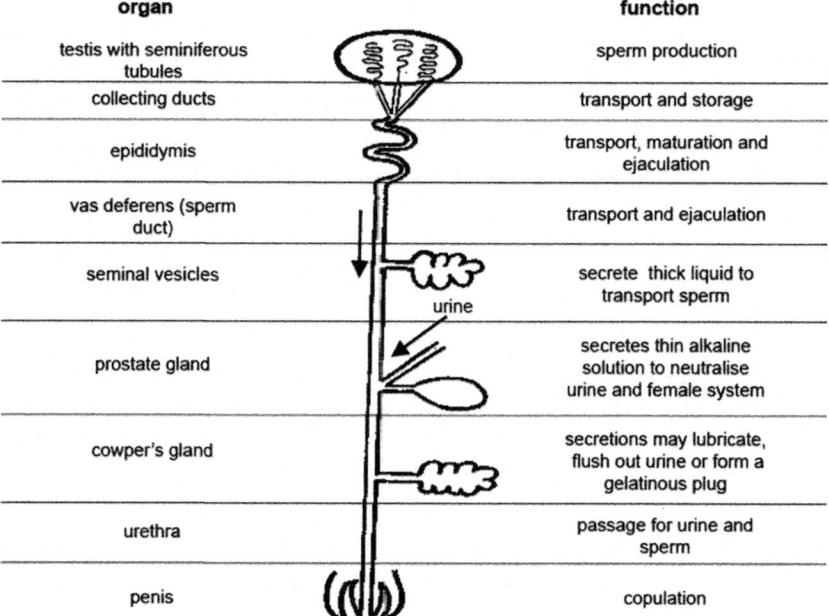

organ	function
testis with seminiferous tubules	sperm production
collecting ducts	transport and storage
epididymis	transport, maturation and ejaculation
vas deferens (sperm duct)	transport and ejaculation
seminal vesicles	secrete thick liquid to transport sperm
prostate gland	secretes thin alkaline solution to neutralise urine and female system
cowper's gland	secretions may lubricate, flush out urine or form a gelatinous plug
urethra	passage for urine and sperm
penis	copulation

Figure 4: Diagram summarising the functions of the male reproductive organs

The Testes

Sperm need temperatures between 2 to 10 degrees Centigrade lower and then the body temperature to develop. This is the reason why the testes are located in a bag of skin called the scrotal sacs (or scrotum) that hangs below the body and where the evaporation of secretions from special glands can further reduce the temperature. In many animals (including humans) the testes descend into the scrotal sacs at birth but in some animals they do not descend until sexual maturity and in others they only descend temporarily during the breeding season. A mature animal in which one or both testes have not descended is called a cryptorchid and is usually infertile.

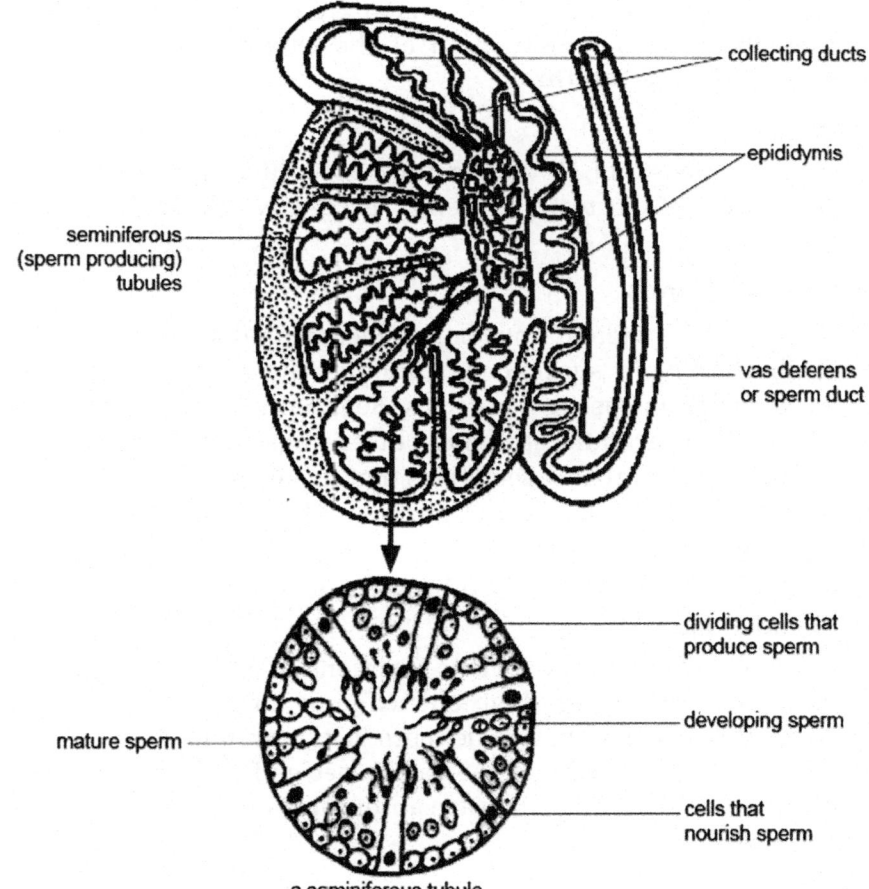

Figure 5: The testis and a magnified seminiferous tubule

The problem of keeping sperm at a low enough temperature is even greater in birds that have a higher body temperature than mammals. For this reason bird's sperm are usually produced at night when the body temperature is lower and the sperm themselves are more resistant to heat.

The testes consist of a mass of coiled tubes (the seminiferous or sperm producing tubules) in which the sperm are formed by meiosis. Cells lying between the seminiferous tubules produce the male sex hormone testosterone. When the sperm are mature they accumulate in the collecting ducts and then pass to the epididymis before moving to the sperm duct or vas deferens. The two sperm ducts join the urethra just below the bladder, which passes through the penis and transports both sperm and urine.

Ejaculation discharges the semen from the erect penis. It is brought about by the contraction of the epididymis, vas deferens, prostate gland and urethra.

Semen

Semen consists of 10% sperm and 90% fluid and as sperm pass down the ducts from testis to penis, (accessory) glands add various secretion.

Accessory Glands

Three different glands may be involved in producing the secretions in which sperm are suspended, although the number and type of glands varies from species to species. Seminal vesicles are important in rats, bulls, boars and stallions but are absent in cats and dogs. When present they produce secretions that make up much of the volume of the semen, and transport and provide nutrients for the sperm.

The prostate gland is important in dogs and humans. It produces an alkaline secretion that neutralizes the acidity of the male urethra and female vagina.

Cowper's glands have various functions in different species. The secretions may lubricate, flush out urine or form a gelatinous plug that traps the semen in the female reproductive system after copulation and prevents other males of the same species fertilizing an already mated female. Cowper's glands are absent in bears, dogs, and aquatic mammals.

The Penis

The penis consists of connective tissue with numerous small blood spaces in it. These fill with blood during sexual excitement causing erection.

Penis Form and Shape

Dogs, bears, seals, bats and rodents have a special bone in the penis which helps maintain the erection. In some animals (e.g. the bull, ram and boar) the penis has an "S" shaped bend that allows it to fold up when not in use. In many animals the shape of the penis is adapted to match that of the vagina. For example, the boar has a corkscrew shaped penis, there is a pronounced twist in bulls' and it is forked in marsupials like the opossum. Some have spines, warts or hooks on them to help keep them in the vagina and copulation may be extended to help retain the semen in the female system. Mating can last up to three hours in minks, and dogs may "knot" or "tie" during mating and can not separate until the erection has subsided.

Sperm

Sperm are made up of three parts: a head consisting mainly of the nucleus, a midpiece containing many mitochondria to provide the energy and a tail that provides propulsion.

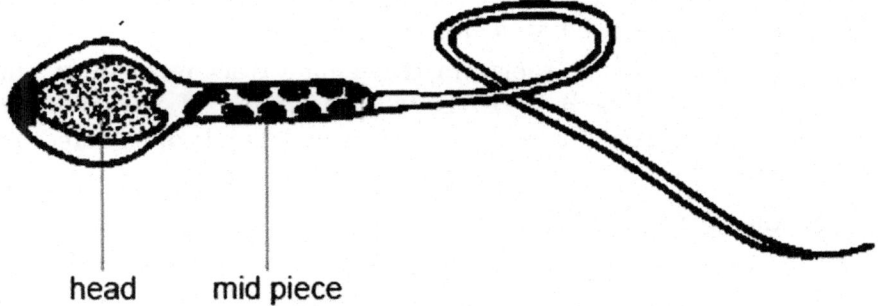

head   mid piece

Figure 6: A sperm

A single ejaculation may contain 2-3 hundred million sperm but even in normal semen as many as 10% of these sperm may be abnormal and infertile. Some may be dead while others are inactive or deformed with double, giant or small heads or tails that are coiled or absent altogether.

When there are too many abnormal sperm or when the sperm concentration is low, the semen may not be able to fertilize an egg and the animal is infertile. Make sure you don't confuse infertility with impotence, which is the inability to copulate successfully.

Sperm do not live forever. They have a definite life span that varies from species to species. They survive for between 20 days (guinea pig) to 60 days (bull) in the epididymis but once ejaculated into the female tract they only live from 12 to 48 hours. When semen is used for artificial insemination, storage under the right conditions can extend the life span of some species.

Artificial Insemination

In many species the male can be artificially stimulated to ejaculate and the semen collected. It can then be diluted, stored and used to inseminate females. For example bull semen can be diluted and stored for up to 3 weeks at room temperature. If mixed with an antifreeze solution and stored in "straws" in liquid nitrogen at minus 79C it will keep for much longer. Unfortunately the semen of chickens, stallions and boars can only be stored for up to 2 days.

Dilution of the semen means that one male can be used to fertilize many more females than would occur under natural conditions. There are also advantages in the male and female not having to make physical contact. It means that owners of females do not have to buy expensive males and the possibility of transmitting sexually transmitted diseases is reduced. Routine examination of the semen for sperm concentration, quality and activity allows only the highest quality semen to be used so a high success rate is ensured.

Since the lifespan of sperm in the female tract is so short and ova only survive from 8 to 10 hours the timing of the artificial insemination is critical. Successful conception depends upon detecting the time that the animal is "on heat" and when ovulation occurs.

The Female Reproductive Organs

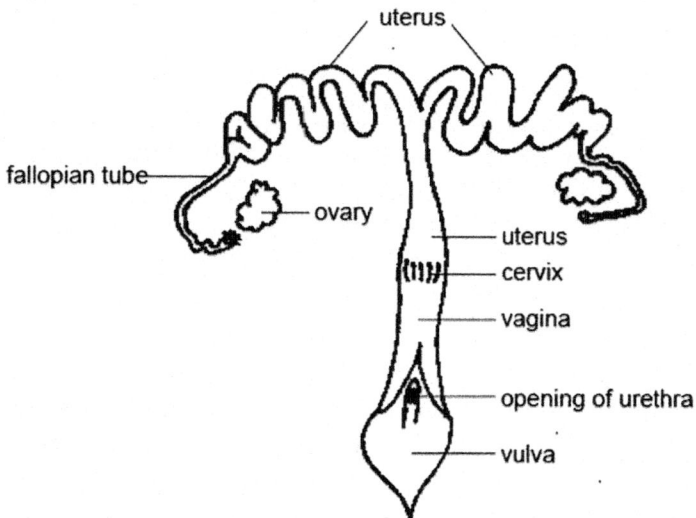

Figure 7: The reproductive system of a female rabbit

The female reproductive system consists of a pair of ovaries that produce egg cells or ova and fallopian tubes where fertilization occurs and which carry the fertilized ovum to the uterus. Growth of the foetus takes place here. The cervix separates the uterus from the vagina or birth canal, where the sperm are deposited.

Note that primates like humans have a uterus with a single compartment but in most mammals the uterus is divided into two separate parts or horns as shown in diagram 13.6.

The Ovaries

Ovaries are small oval organs situated in the abdominal cavity just ventral to the kidneys. Most animals have a pair of ovaries but in birds only the left one is functional to reduce weight.

The ovary consists of an inner region (medulla) and an outer region (cortex) containing egg cells or ova. These are formed in large numbers around the time of birth and start to develop after the animal becomes sexually mature. A cluster of cells called the follicle surrounds and nourishes each ovum.

The Ovarian Cycle

The ovarian cycle refers to the series of changes in the ovary during which the follicle matures, the ovum is shed and the corpus luteum develops.

Numerous undeveloped ovarian follicles are present at birth but they start to mature after sexual maturity. In animals that normally have only one baby at a time only one ovum will mature at once but in litter animals several will. The mature follicle consists of outer cells that provide nourishment. Inside this is a fluid-filled space that contains the ovum.

A mature follicle can be quite large, ranging from a few millimetres in small mammals to the size of a golf ball in large animals. It bulges out from the surface of the ovary before eventually rupturing to release the ovum into the abdominal cavity. Once the ovum has been shed, a blood clot forms in the empty follicle.

This develops into a tissue called the corpus luteum that produces the hormone progesterone. If the animal becomes pregnant the corpus luteum persists, but if there is no pregnancy it degenerates and a new ovarian cycle usually.

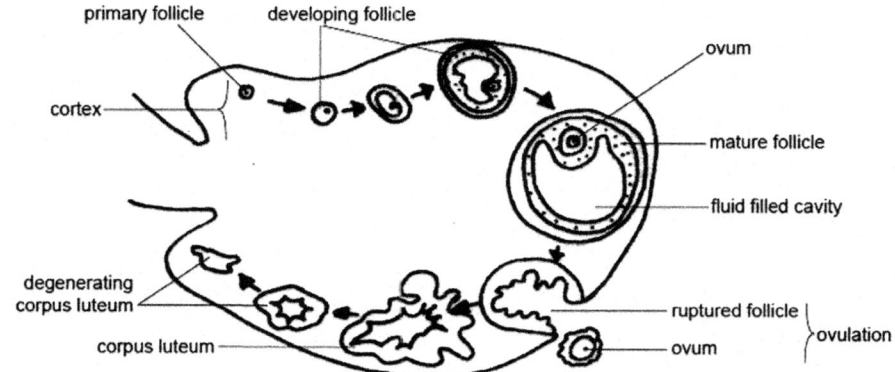

Figure 8: The ovarian cycle showing from the top left clockwise: the maturation of the ovum over time, followed by ovulation and the development of the corpus luteum in the empty follicle

The Ovum

When the ovum is shed the nucleus is in the final stages of meiosis (cell division). It is surrounded by several layers of follicle cells and a tough membrane called the zona pellucida.

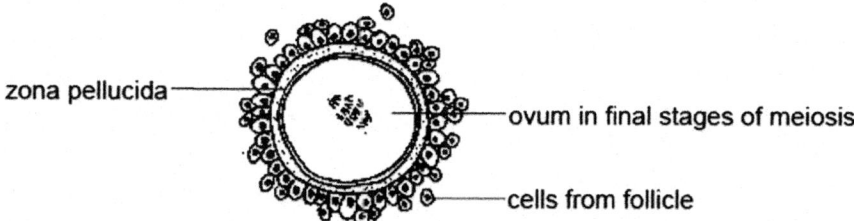

Figure 9: An ovum

The Oestrous Cycle

The oestrous cycle is the sequence of hormonal changes that occurs through the ovarian cycle. These changes influence the behaviour and body changes of the female.

The first hormone involved in the oestrous cycle is follicle stimulating hormone (F.S.H.), secreted by the anterior pituitary gland. It stimulates the follicle to develop. As the follicle matures the outer cells begin to secrete the hormone oestrogen and this stimulates the mammary glands to develop. It also prepares the lining of the uterus to receive a fertilized egg. Ovulation is initiated by a surge of another hormone from the anterior pituitary, luteinising hormone (L.H.). This hormone also influences the development of the corpus luteum, which produces progesterone, a hormone that prepares the lining of the uterus for the fertilized ovum and readies the mammary glands for milk production.

If no pregnancy takes place the corpus luteum shrinks and the production of progesterone decreases. This causes FSH to be produced again and a new oestrous cycle begins.

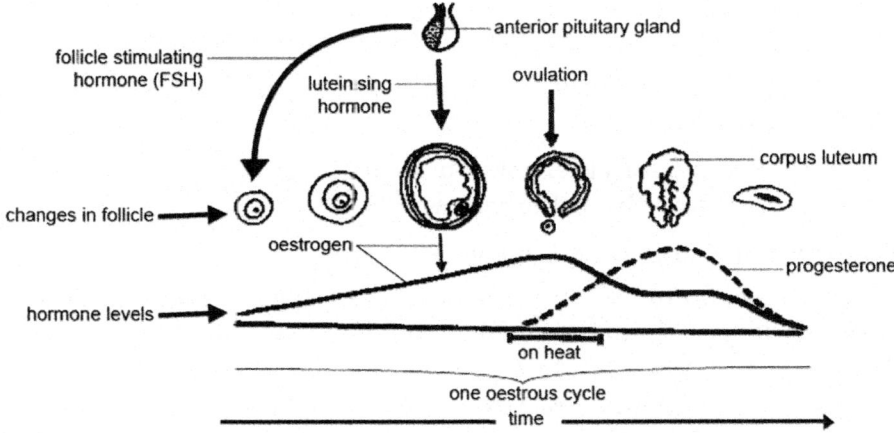

Figure 10: The oestrous cycle

For fertilization of the ovum by the sperm to occur, the female must be receptive to the male at around the time of ovulation. This is when the hormones turn on the signs of "heat", and she is "in season" or "in oestrous". These signs are turned off again at the end of the oestrous cycle.

During the oestrous cycle the lining of the uterus (endometrium) thickens ready for the fertilized ovum to be implanted. If no pregnancy occurs this thickened tissue is absorbed and the next cycle starts. In humans and other higher primates, however, the endometrium is shed as a flow of blood and instead of an oestrous cycle there is a menstrual cycle.

The length of the oestrous cycle varies from species to species. In rats the cycle only lasts 4-5 days and they are sexually receptive for about 14 hours. Dogs have a cycle that lasts 60-70 days and heat lasts 7-9 days and horses have a 21-day cycle and heat lasts an average of 6 days.

Ovulation is spontaneous in most animals but in some, e.g. the cat, and the rabbit, ovulation is stimulated by mating. This is called induced ovulation.

Signs of Oestrous or Heat

- When on heat a bitch has a blood stained discharge from the vulva that changes a little later to a straw coloured one that attracts all the dogs in the neighbourhood.

- Female cats "call" at night, roll and tread the carpet and are generally restless but will "stand" firm when pressure is placed on the pelvic region (this is the lordosis response).

- A female rat shows the lordosis response when on heat. It will "mount" other females and be more active than normal.

- A cow mounts other cows (bulling), bellows, is restless and has a discharge from the vulva.

Breeding Seasons and Breeding Cycles

Only a few animals breed throughout the year. This includes the higher primates (humans, gorillas and chimpanzees etc.), pigs, mice and rabbits. These are known as continuous breeders.

Most other animals restrict reproduction to one or two seasons in the year-seasonal breeders. There are several reasons for this. It means the young can be born at the time (usually spring) when feed is most abundant and temperatures are favourable. It is also sensible to restrict the breeding season because courtship, mating, gestation and the rearing of young can exhaust the energy resources of an animal as well as make them more vulnerable to predators.

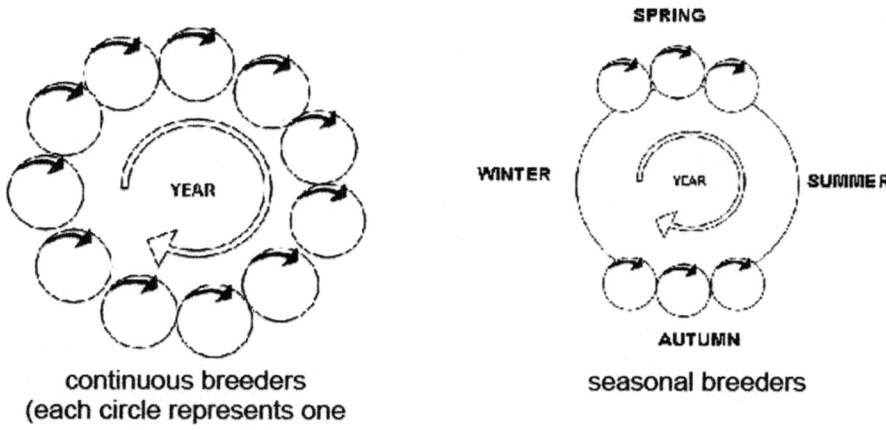

continuous breeders
(each circle represents one
oestrous cycle)

seasonal breeders

Figure 11: Breeding cycles

The timing of the breeding cycle is often determined by day length. For example the shortening day length in autumn will bring sheep and cows into season so the foetus can gestate through the winter and be born in spring. In cats the increasing day length after the winter solstice (shortest day) stimulates breeding.

The number of times an animal comes into season during the year varies, as does the number of oestrous cycles during each season. For example a dog usually has 2-3 seasons per year, each usually consisting of just one oestrous cycle. In contrast ewes usually restrict breeding to one season and can continue to cycle as many as 20 times if they fail to become pregnant.

Fertilization and Implantation

Fertilization: The opening of the fallopian tube lies close to the ovary and after ovulation the ovum is swept into its funnel-like opening and is moved along it by the action of cilia and wave-like contractions of the wall.

Copulation deposits several hundred million sperm in the vagina. They swim through the cervix and uterus to the fallopian tubes moved along by whip-like movements of their tails and contractions of the uterus. During this journey the sperm undergo their final phase of maturation so they are ready to fertilize the ovum by the time they reach it in the upper fallopian tube.

High mortality means only a small proportion of those deposited actually reach the ovum. The sperm attach to the outer zona pellucida and enzymes secreted from a gland in the head of the sperm dissolve this membrane so it can enter. Once one sperm has entered, changes in the zona pellucida prevent further sperm from penetrating. The sperm loses its tail and the two nuclei fuse to form a zygote with the full set of paired chromosomes restored.

Development of the Morula and Blastocyst

As the fertilized egg travels down the fallopian tube it starts to divide by mitosis. First two cells are formed and then four, eight, sixteen, etc. until there is a solid ball of cells. This is called a morula. As division continues a hollow ball of cells develops. This is a blastocyst.

Implantation

Implantation involves the blastocyst attaching to, and in some species, completely sinking into the wall of the uterus.

Pregnancy

The Placenta and Foetal Membranes: As the embryo increases in size, the placenta, umbilical cord and foetal membranes (often known collectively as the placenta) develop to provide it with nutrients and remove waste products. In later stages of development the embryo becomes known as a foetus.

The placenta is the organ that attaches the foetus to the wall of the uterus. In it the blood of the foetus and mother flow close to each other but never mix. The closeness of the maternal and foetal blood systems allows diffusion between them. Oxygen and nutrients diffuse from the mother's blood into that of the foetus and carbon dioxide and excretory products diffuse in the other direction. Most maternal hormones (except adrenaline), antibodies, almost all drugs (including alcohol), lead and DDT also pass across the placenta. However, it protects the foetus from infection with bacteria and most viruses.

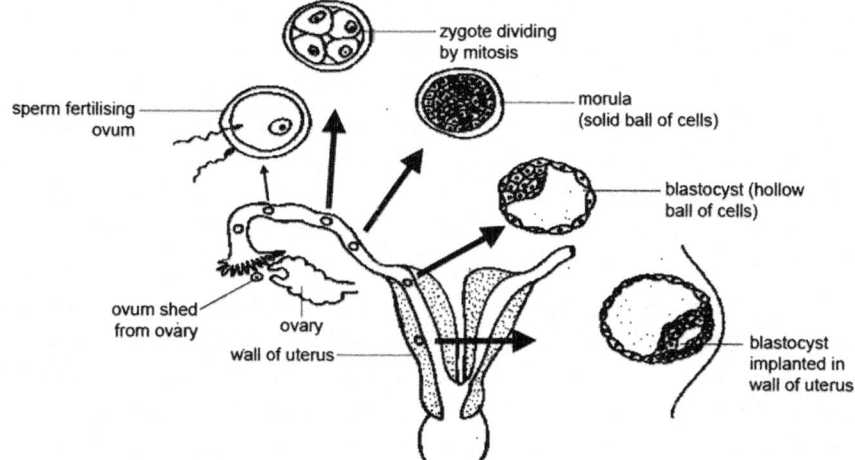

Figure 12: Development and implantation of the embryo

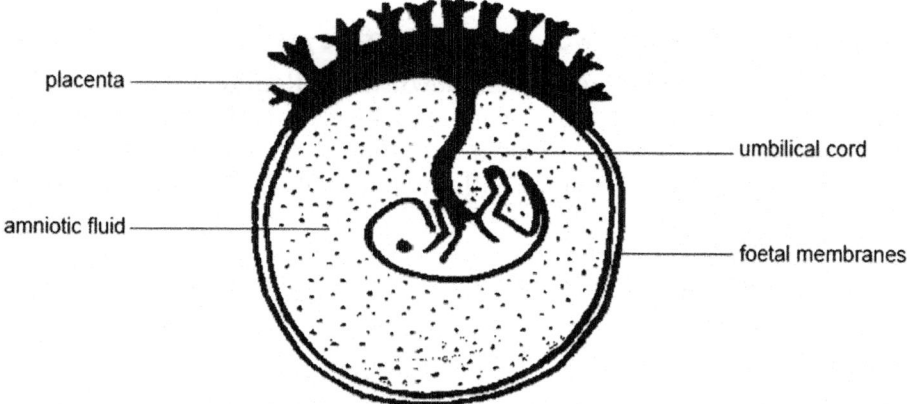

Figure 13: The foetus and placenta

The foetus is attached to the placenta by the umbilical cord. It contains arteries that carry blood to the placenta and a vein that returns blood to the foetus. The developing foetus becomes surrounded by membranes. These enclose the amniotic fluid that protects the foetus from knocks and other trauma.

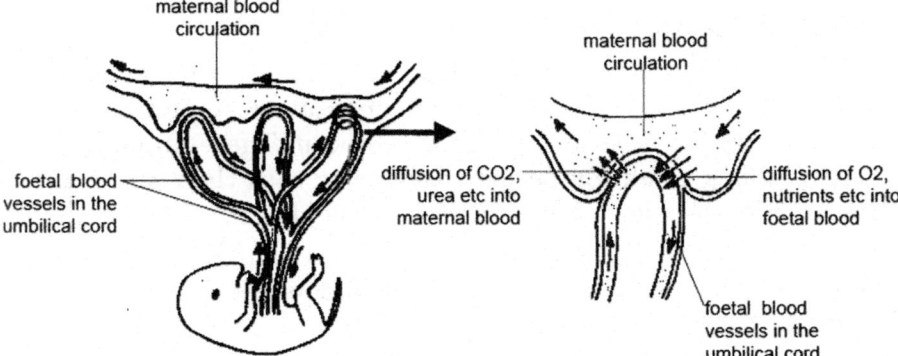

Figure 14: Maternal and foetal blood flow in the placenta

Hormones During Pregnancy

The corpus luteum continues to secrete progesterone and oestrogen during pregnancy. These maintain the lining of the uterus and prepare the mammary glands for milk secretion. Later in the pregnancy the placenta itself takes over the secretion of these hormones.

Chorionic gonadotrophin is another hormone secreted by the placenta and placental membranes. It prevents uterine contractions before labour and prepares the mammary glands for lactation. Towards the end of pregnancy the placenta and ovaries secrete relaxin, a hormone that eases the joint between the two parts of the pelvis and helps dilate the cervix ready for birth.

Pregnancy Testing

The easiest method of pregnancy detection is ultrasound which is noninvasive and very reliable Later in gestation pregnancy can be detected by taking x-rays.

In dogs and cats a blood test can be used to detect the hormone relaxin.

In mares and cows palpation of the uterus via the rectum is the classic way to determine pregnancy. It can also be done by detecting the hormones progesterone or equine chorionic gonagotrophin (eCG) in the urine. A new sensitive test measures the amount of the hormone, oestrone sulphate, present in a sample of faeces. The hormone is produced by the foal and placenta, and is only present when there is a living foal.

In most animals, once pregnancy is advanced, there is a window of time during which an experienced veterinarian can determine pregnancy by feeling the abdomen.

Gestation Period

The young of many animals (e.g. pigs, horses and elephants) are born at an advanced state of development, able to stand and even run to escape predators soon after they are born. These animals have a relatively long gestation period that varies with their size e.g. from 114 days in the pig to 640 days in the elephant.

In contrast, cats, dogs, mice, rabbits and higher primates are relatively immature when born and totally dependent on their parents for survival. Their gestation period is shorter and varies from 25 days in the mouse to 31 days in rabbits and 258 days in the gorilla.

The babies of marsupials are born at an extremely immature stage and migrate to the pouch where they attach to a teat to complete their development. Kangaroo joeys, for example, are born 33 days after conception and opossums after only 8 days.

Birth

Signs of Imminent Birth: As the pregnancy continues, the mammary glands enlarge and may secrete a milky substance a few days before birth occurs. The vulva may swell and produce thick mucus and there is sometimes a visible change in the position of the foetus. Just before birth the mother often becomes restless, lying down and getting up frequently. Many animals seek a secluded place where they may build a nest in which to give birth.

Labour

Labour involves waves of uterine contractions that press the foetus against the cervix causing it to dilate. The foetus is then pushed through the cervix and along the vagina before being delivered. In the final stage of labour the placenta or "afterbirth" is expelled.

Adaptations of the Foetus to Life Outside the Uterus

The foetus grows in the watery, protected environment of the uterus where the mother supplies oxygen and nutrients, and waste products pass to her blood circulation for excretion. Once the baby animal is born it must start to breathe for itself, digest food and excrete its own waste. To allow these functions to occur blood is re-routed to the lungs and the glands associated with the gut start to secrete. Note that newborn animals can not control their own body temperature. They need to be kept warm by the mother, littermates and insulating nest materials.

Milk Production

Cows, manatees and primates have two mammary glands but animals like pigs that give birth to large litters may have as many as 12 pairs. Ducts from the gland lead to a nipple or teat and there may be a sinus where the milk collects before being suckled.

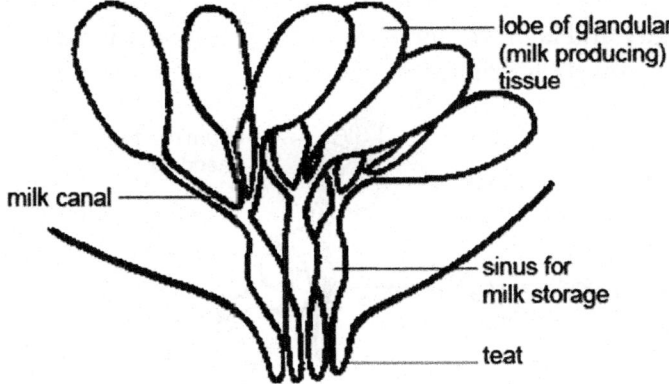

Figure 15: A mammary gland

The hormones oestrogen and progesterone stimulate the mammary glands to develop and prolactin promotes the secretion of the milk. Oxytocin from the pituitary gland releases the milk when the baby suckles. The first milk is called colostrum.

It is a rich in nutrients and contains protective antibodies from the mother. Milk contains fat, protein and milk sugar as well as vitamins and most minerals although it contains little iron. Its actual composition varies widely from species to species. For example whale's and seal's milk has twelve times more fat and four times more protein than cow's milk. Cow's milk has far less protein in it than cat's or dog's milk. This is why orphan kittens and puppies cannot be fed cow's milk.

Reproduction In Birds

Male birds have testes and sperm ducts and male swans, ducks, geese and ostriches have a penis. However, most birds make do with a small amount of erectile tissue known as a papilla. To reduce weight for flight most female birds only have one ovary - usually the left, which produces extremely yolky eggs.

The eggs are fertilized in the upper part of the oviduct (equivalent to the fallopian tube and uterus of mammals) and as they pass down it albumin (the white of the egg), the membrane beneath the shell and the shell are laid down over the yolk. Finally the egg is covered in a layer of mucus to help the bird lay it.

Most birds lay their eggs in a nest and the hen sits on them until they hatch. Ducklings and chicks are relatively well developed when they hatch and able to forage for their own food. Most other nestlings need their parents to keep them warm, clean and fed. Young birds grow rapidly and have voracious appetites that may involve the parents making up to 1000 trips a day to supply their need for food.

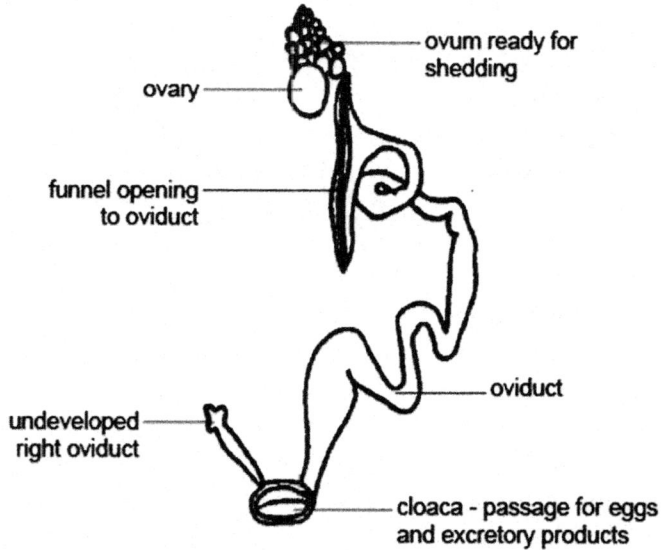

Figure 16: Female reproductive organs of a bird

- Haploid gametes (sperm and ova) are produced by meiosis in the gonads (testes and ovaries).

- Fertilization involves the fusing of the gametes to form a diploid zygote.

- The male reproductive system consists of a pair of testes that produce sperm (or spermatozoa), ducts that transport the sperm to the penis and glands that add secretions to the sperm to make semen.

- Sperm are produced in the seminiferous tubules, are stored in the epididymis and travel via the vas deferens or sperm duct to the junction of the bladder and the urethra where various accessory glands add secretions. The fluid is now called semen and is ejaculated into the female system down the urethra that runs down the centre of the penis.

- Sperm consist of a head, a midpiece and a tail.

- Infertility is the inability of sperm to fertilize an egg while impotence is the inability to copulate successfully.

- The female reproductive system consists of a pair of ovaries that produce ova and fallopian tubes where fertilization occurs and which carry the fertilized ovum to the uterus. Growth of the foetus takes place here. The cervix separates the uterus from the vagina, the birth canal and where the sperm are deposited.

- The ovarian cycle refers to the series of changes in the ovary during which the follicle matures, the ovum is shed and the corpus luteum develops.

- The oestrous cycle is the sequence of hormonal changes that occurs through the ovarian cycle. It is initiated by the secretion of follicle stimulating hormone (F.S.H.), by the anterior pituitary gland which stimulates the follicle to develop. The follicle secretes oestrogen which stimulates mammary gland development. Luteinising hormone (L.H.) from the anterior pituitary initiates ovulation and stimulates the corpus luteum to develop. The corpus luteum produces progesterone that prepares the lining of the uterus for the fertilized ovum.

- Signs of oestrous or heat differ. A bitch has a blood stained discharge, female cats and rats are restless and show the lordosis response, while cows mount other cows, bellow and have a discharge from the vulva.

- After fertilization in the fallopian tube the zygote divides over and over by mitosis to become a ball of cells called a morula. Division continues to form a hollow ball of cells called the blastocyst. This is the stage that implants in the uterus.

- The placenta, umbilical cord and foetal membranes (known as the placenta) protect and provide the developing foetus with nutrients and remove waste products.

Animals and their Environment

A first step to understanding individual animals, and in turn populations of animals, is to understand the relationship they have with their environment. The environment in which an animal lives is referred to as its habitat. A habitat includes both biotic (living) and abiotic (non-living) compents of the animals environment. Abiotic components of an animal's environment include a huge range of characteristics, examples of which are:

- temperature
- humidity
- oxygen
- wind
- soil composition
- day length
- elevation.

Biotic components of an animal's environment include such things as:

- plant matter
- predators
- parasites
- competitors
- individuals of the same species.

Animals require energy to support the processes of life: movement, foraging, digestion, reproduction, growth, work. Organisms can be categorised into one of the following groups:

- autotroph - an organism that obtains energy from sunlight (in the case of green plants) or inorganic compounds (in the case of sulfur bacteria)
- heterotroph - an organism that use organic materials as a source of energy.

Animals are heterotrophs, obtaining their energy from the ingestion of other organisms. When resources are scarce or environmental conditions limit the ability of animals to obtain food or go about their normal activities, animals' metabolic activity may decrease to conserve energy until better conditions prevail. The different types of metabolic dormancy or responses include:

- torpor - a time of decreased metabolism and reduced body temperature in daily activity cycles
- hibernation - a time of decreased metabolism and reduced body temperature that may last weeks or months

- winter sleep - periods of inactivity during which body temperature does not fall substantially and from which animals can be awakened and become active quickly
- aestivation - a period of inactivity in animals that must sustain extended periods of drying

Environmental characteristics (temperature, moisture, food availability, and so on) vary over time and location and animals are adapted to certain range of values for each characteristic. The range of an environmental characteristic to which an animal is adapted is called its tolerance range for that characteristic. Within an animal's tolerance range is an optimal range of values at which the animal is most successful. Sometimes, in response to prolonged change in environmental characteristic, animals' physiology adjusts to accomodate the change in its environment, and in doing so, its tolerance range shifts. This shift in tolerance range that the animal experiences in response to an changed environment, is called acclimation.

Fungi

Fungi are classified by the methods of sexual reproduction they employ. The outcome of sexual reproduction most often is the production of resting spores that are used to survive inclement times and to spread. There are typically three phases in the sexual reproduction of fungi: plasmogamy, karyogamy and meiosis.

Insects

Insect species make up more than two-thirds of all extant animal species, and most insect species use sex for reproduction, though some species are facultatively parthenogenetic. Many species have sexual dimorphism, while in others the sexes look nearly identical. Typically they have two sexes with males producing spermatozoa and females ova. The ova develop into eggs that have a covering called the chorion, which forms before internal fertilization. Insects have very diverse mating and reproductive strategies most often resulting in the male depositing spermatophore within the female, which stores the sperm until she is ready for egg fertilization. After fertilization, and the formation of a zygote, and varying degrees of development; the eggs are deposited outside the female in many species, or in some, they develop further within the female and live born offspring are produced.

Mammals

There are three extant kinds of mammals: Monotremes, Placentals and Marsupials, all with internal fertilization. In placental mammals, offspring are born as juveniles: complete animals with the sex organs present although not reproductively functional. After several months or years, the sex organs develop further to maturity and the animal becomes sexually mature. Most female mammals are only fertile during certain periods during their estrous cycle, at which point they are ready to mate. Individual male and female mammals meet and carry out copulation. For most mammals, males and females exchange sexual partners throughout their adult lives.

Reproductive System of Gastropods

The reproductive system of gastropods (slugs and snails) varies greatly from one group to another within this very large and diverse taxonomic class of animals. Their reproductive strategies also vary greatly.

In many marine gastropods there are separate sexes (male and female); most terrestrial gastropods however are hermaphrodites. Courtship is a part of mating behaviour in some gastropods. In some families of pulmonate land snails, one unusual feature of the reproductive system and reproductive behaviour is the creation and utilisation of love darts.

Reproduction in Marine Gastropods

Separate Sexes: In many taxonomic groups of marine gastropods, there are separate sexes (i.e. they are dioecious).

The great majority of species in some of the main gastropod clades have separate sexes. This is true in most of the Patellogastropoda, Vetigastropoda, Cocculiniformia, Neritimorpha, and Caenogastropoda.

Protandrous Sequential Hermaphrodites

Within the clade Littorinimorpha however, the superfamily Calyptraeoidea are protandrous sequential hermaphrodites. Protandry means that the individuals first become male, and then later on become female.

Simultaneous Hermaphrodites

Within the main clade Heterobranchia, the informal group Opisthobranchia are simultaneous hermaphrodites (they have both sets of reproductive organs within one individual at the same time).

Bibliography

Adams, Carol.: *The Sexual Politics of Meat: A Feminist-Vegetarian Critical Theory,* Continuum, New York, 1991.

Agamben, Giorgio: *The Open: Man and Animal,* Stanford University Press, UK, 2004.

Alexander, Shana: *The Astonishing Elephant,* Random House, New York, 2000.

Alger, Janet and Steven Alger: *Cat Culture: The Social World of a Cat Shelter,* Temple University Press, Philadelphia, 2003.

Allen, Karen Miller: *The Human-Animal Bond: An Annotated Bibliography,* Scarecrow Press, Metuchen, NJ, 1985.

Allister, Mark Christopher: *Eco-man: New Perspectives on Masculinity and Nature,* University of Virginia Press, Charlottesville, 2004.

Anderson, R. S.: *Pet Animals and Society,* Bailliere Tindall, London, 1984.

Armstrong, Susan J. and Richard G. Botzler: *The Animal Ethics Reader,* Routledge, New York, 2003.

Baker, Steve: *The Postmodern Animal,* Reaktion Books, London, 2000.

Baldick, Julian: *Animals and Shaman: Ancient Religions of Central Asia,* New York University Press, New York, 2000.

Baratay, Eric.: *Zoo: A History of Zoological Gardens in the West,* Reaktion, London, 2002.

Barton, M.: *Animal Rights,* Watts, London, 1987.

Bauston, Gene: *Battered Birds, Crated Herds: How We Treat the Animals We Eat,* Farm Sanctuary, Watkins Glen, NY, 1996.

Beck, Alan M. and Aaron Katcher: *Between Pets and People: The Importance of Animal Companionship,* Putnam, New York, 1983.

Bekoff, Marc and Jessica Pierce: *Wild Justice: The Moral Lives of Animals,* The University of Chicago Press, Chicago, 2009.

Bekoff, Marc: *The Smile of a Dolphin: Remarkable Accounts of Animal Emotions,* Discovery Books, New York, 2000.

Benton, Ted: *Natural Relations: Ecology, Animal Rights and Social Justice,* Verso, London, 1993.

Boehrer, Bruce: *A Cultural History of Animals in the Renaissance,* Oxford, UK: Berg Publishers, 2007.

Bright, M.: *Animal Language,* BBC Publications, London, 1984.

Brown, L.: *Cruelty to Animals: The Moral Debt,* MacMillan, London, 1988.

Brunner, Bernd: *Bears: A Brief History,* Yale University Press, New Haven, 2007.

Budiansky, S.: *The Covenant of the Wild: Why Animals Chose Domestication,* Morrow, New York, 1992.

Burt, Jonathan: *Rat,* Reaktion, London, 2004.

Calarco, Matthew, and Peter Atterton: *Animal Philosophy: Essential Readings in Continental Thought,* Continuum, London, 2004.

Carlson, Laurie: *Cattle: An Informal Social History,* Ivan R. Dee, Chicago, IL, 2001.

Carter, Paul: *Parrot,* Reaktion, London, 2005.

Chris, Cynthia: *Watching Wildlife,* University of Minnesota Press, Minneapolis, 2006.

Clark, S. and S. Lyster: *Animals and Their Moral Standing,* Routledge, London, 1997.

Clark, Stephen: *The Nature of the Beast.* Oxford University Press, New York, 1984.

Clutton-Brock, Juliet: *Domesticated Animals from Early Times,* University of Texas Press, Austin, TX, 1981.

Copeland, Marion W.: *Cockroach.* Reaktion, London, 2003.

Coren, Stanley: *The Pawprints of History: Dogs and the Course of Human Events,* Free Press, New York, 2002.

Crist, Eileen: *Images of Animals: Anthropomorphism and Animal Mind,* Temple University Press, Philadelphia, 1999.

DeGrazia, David: *Taking Animals Seriously: Mental Life and Moral Status,* Cambridge, New York, 1996.

Derr, Mark: *Dog's Best Friend: Annals of the Dog-human Relationship,* University of Chicago Press, Chicago, 2004.

Dolins, Francine: *Attitudes to Animals: Views on Animal Welfare,* Cambridge University Press, Cambridge, 1999.

Fine, Aubrey H.: *Handbook on Animal-Assisted Therapy: Theoretical Foundations and Guidelines for Practice,* Academic Press, UK, 2006.

Fitzgerald, Amy J.: *Animal Abuse and Family Violence: Researching the Interrelationships of Abusive Power,* Edwin Mellen Press, Lewiston, New York, 2005.

Fogle, Bruce: *Pets and Their People,* The Viking Press, New York, 1983.

Fox, Michael Allen: *The Case for Animal Experimentation,* University of California Press, Berkeley, CA, 1986.

Friend, Tim: *Animal Talk: Breaking the Codes of Animal Language,* Free Press, New York, 2004.

Fudge, Erica: *Brutal Reasoning: Animals, Rationality and Humanity in Early Modern England*, Cornell University Press, Ithaca, 2006.

Gallistel, C.R.: *Animal Cognition*, MIT Press, Cambridge, 1992.

Gates, P.: *Animal Communication*, Cambridge University Press, Cambridge, 1997.

Harbolt-Bosco, Tami: *Bridging the Bond: The Cultural Construction of the Shelter Pig*, Purdue University Press, 2003.

Harris, Marvin: *The Sacred Cow and the Abominable Pig: Riddles of Food and Culture*, Touchstone Books, New York, 1987.

Hearne, Vicki: *Animal Happiness*, HarperCollins, New York, 1994.

Hyland, Ann: *The Horse in the Ancient World*, Praeger, Westport, CT, 2003.

Jones, Susan D.: *Valuing Animals: Veterinarians and Their Patients in Modern America*, Johns Hopkins University Press, Baltimore, MD, 2003.

Katz, Jon: *The New Work of Dogs: Tending to Life, Love, and Family*, Villard, New York, 2003.

Kerasote, Ted: *Bloodties: Nature, Culture, and the Hunt*, Kodansha, New York, 1993.

Langley , Gill: *Animal Experimentation: The Consensus Changes*, Chapman and Hall, New York, 1989.

Lilequist , J.: *Forbidden History, the State, Society and the Regulation of Sexuality in Modern Europe*, Chicago University Press, Chicago, 1992.

MacDonald, Helen: *Falcon*, Reaktion, London, 2005.

Manning, Aubrey, and James Serpell: *Animals and Human Society*, Routledge, New York, 1994.

Mason, Jim and Peter Singer: *Animal Factories*, Crown Publishers, New York, 1980.

Masson, Jeffrey Moussaieff: *Dogs Never Lie About Love: Reflections on the Emotional World of Dogs*, Crown Publishers, New York, 1997.

McHugh, Susan: *Dog*, Reaktion, London, 2004.

Mitchell, W.J.T.: *The Last Dinosaur Book: The Life and Times of a Cultural Icon*, University of Chicago Press, Chicago and London, 1998.

Morris, Desmond: *The Naked Ape*, McGraw-Hill, New York, 1967.

Morton, Eugene S.: *Animal Talk: Science and the Voices of Nature*, Random House, New York, 1992.

Noske, Barbara: *Humans and Other Animals: Beyond the Boundaries of Anthropology*, Pluto Press, London, 1989.

Phillips, Ian: *Lost: Lost and Found Pet Posters from Around the World*, Princeton Architectural Press, New York, 2002.

Porter, V.: *Faithful Companions: The Alliance of Man and Dog*, Methuen, London, 1989.

Rifkin, Jeremy: *Beyond Beef: The Rise and Fall of the Cattle Culture*, Dutton, New York, 1992.

Ristau, C.: *Cognitive Ethology: The Minds of Other Animals*, Lawrence Erlbaum, Hillsdale, NJ, 1990.

Robbins, Louise E.: *Elephant Slaves and Pampered Parrots: Exotic Animals in Eighteenth-Century Paris*, Johns Hopkins University Press, Baltimore, 2002.

Roger, J.: *Buffon: A Life in Natural History*, Cornell University Press, Ithaca , NY, 1997.

Rogers, Katherine: *The Cat and the Human Imagination*, The University of Michigan Press, Ann Arbor, 2001.

Salem, Harry: *Animal Test Alternatives: Refinement, Reduction, Replacement*, M. Dekker, New York, 1995.

Sanders, Clinton: *Understanding Dogs*, Temple University Press, Philadelphia, 1999.

Sax, Boria: *Crow*, Reaktion Books, London, 2003.

Sharpe, Robert: *Science on Trial: The Human Cost of Animal Experiments*, Awareness Books, Sheffield, UK, 1994.

Sleigh, Charlotte: *Ant,* Reaktion, London, 2003.

Stott, Rebecca: *Oyster,* Reaktion, London, 2004.

Stutesman, Drake: *Snake,* Reaktion, London, 2005.

Swan, James: *In Defense of Hunting*, HarperCollins, New York, 1995.

Tabor, Roger K.: *The Wild Life of the Domestic Cat,* Arrow Books, London, 1983.

Taylor, R.E. and R. Bogart: *Scientific Farm Animal Production: An Introduction to Animal Science*, MacMillan, New York, 1988.

Thomas, Elizabeth Marshall: *The Tribe of the Tiger,* Simon and Schuster, New York, 1994.

Topsells, E.: *History of Four-footed Beast,* Nelson Hall, Chicago, 1981.

Ucko, Peter J. and G.W. Dimbleby: *The Domestication and Exploitation of Plants and Animals*, Aldine Publishing Company, Chicago, IL, 1969.

Willis, R.G.: *Signifying Animals: Human Meaning in the Natural World,* Unwin Hyman, London, 1990.

Yarri, Donna: *The Ethics of Animal Experimentation: A Critical Analysis And Constructive Christian Proposal*, Oxford University Press, Oxford, 2005.

Zeuner, Frederick E.: *A History of Domesticated Animals*, Hutchinson, London, 1963.

Index

❑❑❑